Studies in Systems, Decision and Control

Volume 370

Series Editor

Janusz Kacprzyk, Systems Research Institute, Polish Academy of Sciences, Warsaw, Poland

The series "Studies in Systems, Decision and Control" (SSDC) covers both new developments and advances, as well as the state of the art, in the various areas of broadly perceived systems, decision making and control–quickly, up to date and with a high quality. The intent is to cover the theory, applications, and perspectives on the state of the art and future developments relevant to systems, decision making, control, complex processes and related areas, as embedded in the fields of engineering, computer science, physics, economics, social and life sciences, as well as the paradigms and methodologies behind them. The series contains monographs, textbooks, lecture notes and edited volumes in systems, decision making and control spanning the areas of Cyber-Physical Systems, Autonomous Systems, Sensor Networks, Control Systems, Energy Systems, Automotive Systems, Biological Systems, Vehicular Networking and Connected Vehicles, Aerospace Systems, Automation, Manufacturing, Smart Grids, Nonlinear Systems, Power Systems, Robotics, Social Systems, Economic Systems and other. Of particular value to both the contributors and the readership are the short publication timeframe and the world-wide distribution and exposure which enable both a wide and rapid dissemination of research output.

Indexed by SCOPUS, DBLP, WTI Frankfurt eG, zbMATH, SCImago.

All books published in the series are submitted for consideration in Web of Science.

More information about this series at http://www.springer.com/series/13304

Benedito Medeiros Neto · Inês Amaral ·
George Ghinea
Editors

Digital Convergence in Contemporary Newsrooms

Media Innovation, Content Adaptation,
Digital Transformation, and Cyber Journalism

 Springer

Editors
Benedito Medeiros Neto
University of Brasília
Brasilia, Brazil

Inês Amaral
Faculty of Arts and Humanities
University of Coimbra
Coimbra, Portugal

George Ghinea
College of Engineering, Design
and Physical Sciences
Brunel University London
Uxbridge, UK

ISSN 2198-4182   ISSN 2198-4190   (electronic)
Studies in Systems, Decision and Control
ISBN 978-3-030-74430-4   ISBN 978-3-030-74428-1   (eBook)
https://doi.org/10.1007/978-3-030-74428-1

This Springer imprint is published by the registered company Springer Nature Switzerland AG
The registered company address is: Gewerbestrasse 11, 6330 Cham, Switzerland

Nous autres écrivains (...) ne disposons du pouvoir politique ni des forces armées ni d'argent. (...) Nous ne disposons donc que du langage et,
parfois, de l'enseignement. Nous ne pouvons donc
que travailler à long terme.

Exactement dans celui du Grand Récit.[6]

We writers (...) do not have political power nor armed forces nor money. (...) Thus, we have only language, and sometimes, teaching.
Thus, we can only work in the long term.
Precisely the term of the Great Narrative.

---

[6]Michel Serres in *L'Incandescent*, p. 405. Free translation into English.

Journalists and Health

Professionals Struggle for Life

We dedicate this book to journalists who died of COVID-19[1] as well as to health professionals—nurses, doctors, medics, hospital personnel, and public health units, who constitute an immense chain of expertise, fighting to save lives often without the necessary protection equipment. These people are in a much larger number than journalists; thus, they have many more dead.

The most advanced technology and the highest science do not seem able to contain a virus that spreads at unprecedented speed. Meanwhile, various countries elect governments with a mentality akin to the Middle Ages. Once in power, they mock the disease and act against the same democracy that elected them. They ignore science. They insult intelligence.

However, they do not despise the knowledge of Information and Communication

---

[1]Two hundreds and ten journalists in 38 countries between March and mid-July 2020. See Press Emblem Campaign.

Technologies (ICT) to reach their power-thirsty goals. Understood here as a civilizational advancement and as the search for knowledge accessibility, the ICT are used to identify the most resentful segments of the population, which are also the most vulnerable to hate speech. The same social networks used by people to communicate and to become authors, not just spectators of the media universe, are also weaponized by political and economic groups to pursue the opposite direction, often times criminally. They poke wounds and stimulate resentment and frustration, in order to have their preys act against life and democracy.

In the bosom of formal democracies, totalitarian alternatives thrive.

Journalists and health professionals are not the means for such ideologies. They want safer, well-informed populations. But in the USA, India, Brazil, and many African, European, and Asian countries, these professionals face on a daily basis the double threats of virus and disinformation. They became a target for the hate speech that poisons social networks. Were not the Internet and social networks a possibility for liberation, for increased communication, and for expanding democracy? Behold the paradox Earth inhabitants have to deal with.

We will work toward having more and more humans fighting for life in our planet. May scientists, researchers, journalists, health professionals, and other professionals too be

*the bearers of the highest ethical values. May
this book be part of a path in that direction.
Many will join in this effort.*

Fred Ghedini
*Brazilian Journalist, President of Associação
Profissão Jornalista (APJor)*

Foreword: What Aeschylus Can Teach Us About Disinformation and the Importance of Journalism

When looking for some additional readings for this preface, I turned to Aeschylus as the perfect option to illustrate how deception and manipulation are ancient, especially in times of conflict. Personally, I believe that problematizing the importance of journalism in the twenty-first century is first and foremost thinking about the value of truth in a technologically mediated world. So, I thought that a good opening to this text could simultaneously lead the way and illustrate how far back dates the manipulation of truth. I'm always skeptical when I read that the first manifestation of fake news occurred in this or that precise moment and context. It surely must be difficult to identify that. But I thought that it would be safe to bring the words of Aeschylus, the Greek playwriter who lived around 500 years prior to the birth of Christ. "In war, truth is the first casualty," said Aeschylus. I had quoted it more than once. I googled it once again and quickly found it, which reassured me: It is true and anyone can check it online. However, I wanted to be more accurate and went looking for the source of the aphorism. Aeschylus probably said it on a speech or wrote it in one of its plays. My search for the moment and place of that statement showed me that it was not just him, the father of tragedy as he is often called, who was presented as the author of the adage. It was also attributed to others. So, I investigated deeper and even sought for guidance from a scholar in classic Greek culture.

Eventually, two online pages put an end to my quest: a "Quote investigator"[2] and a "Wikiquote"[3] guaranteed that it was misattributed. There is no evidence that Aeschylus wrote it or ever verbalized it. However, it Is widely spread on the web. Gaetani describes precisely this: "during my research I have contacted many bloggers, asking them where Camus should have written/said this or that; their answer was always the same: «check it on Google». Indeed, their reasoning was simple but tremendously naïve: if a quote is reported by so many people—millions of references in some cases—the author of this quote 'must' be Albert Camus"

---

[2]https://quoteinvestigator.com/2020/04/11/casualty/ on 20 January 2021.

[3]https://en.wikiquote.org/wiki/Aeschylus.

(2015, p. 41). This episode encapsulates various aspects that substantiate my view of why we should be discussing what is happening to journalism today.

Journalism has been facing dramatic obstacles and shifts since the twentieth century. However, the nature of these problems has been changing over time (Nielsen 2016; Pickard 2011). What we face today can be analyzed mainly from three angles: "first, an economic crisis concerning the very existence of the news media industry that underwrites journalism as an occupation, a form of salaried employment; second, a professional crisis concerning the demarcation of journalism itself, its separation from other kinds of work; and third, a crisis of confidence concerning the relations between journalism and the people who make up the public that journalism claims and aims to serve (Nielsen 2016, p. 77). Stemming from this threefold view of the crisis of journalism, I intend to analyze how the current techno-social, economic and political contexts pose enormous threats to journalism and democracy—after all, they are tangled together, and how people, more than ever, need trustworthy information.

Despite the promise of a more open and participatory system that cyberoptimists announced in the 1990s, what we see today is the permanence of mainstream media as the dominant actors. There is an immense multitude of "produsers" (Bruns 2007) and a new breed of digital influencers. Nevertheless, the dominance of mainstream legacy media perpetuates what James Curran described as a "closeness between the media and the Establishment" (2019, p. 191), that can undermine the way journalism is seen by the public opinion. During the twentieth century, journalism was able to create an image of true value, a non-partisan means that allowed people to get the information they needed to interpret the world and govern themselves. Nevertheless, journalism is/was never bias-proof. It depends on choices and priorities. It delivers information, a public good that supports democracy, but it never ceases being a business.[4] In the USA, for example, the clashes between profit and news values were increasingly won by the first, during the 1980s and the 1990s: "profit maximization came to dominate all other considerations" (Benson 2018, p. 1061). At the turn of the millennium, right before the Lehman Brothers collapse, American journalism was a product of its time: "a political economic analysis stresses that the reasons for lousy journalism stem not from morally bankrupt or untalented journalists, but from a structure that makes such journalism the rational result of its operations" wrote McChesney in 2003 (2003, p. 324).

What we see today is also a product of our time. The kind of journalism that we have results from the answers (some of which desperate) that were given by the media outlets to the dramatic changes in the media ecosystem, either caused financial restrictions, by the rise of new players, or by new consumption habits, among other reasons. One of the most relevant causes is the end of the business model that comfortably characterized journalism during the mass media monopoly (Pickard 2011). A few years ago, Benito-Ruiz (2009) coined the term "infoxication 2.0" to refer to a time which we could describe as highly saturated. A time that is

---

[4]In this text, the essence of the public service media or other forms, like community media is not discussed.

characterized by a permanent online stimulus and by the pressure of an ever-present technology. The fierce struggle to get more clicks, which is arguably the current measure of success for media outlets, is fought with the weapons of "Infoxication 2.0": promoting infotainment, exploiting highly attractive and sensational topics, worrying less with deontology and focusing more in viralization strategies (Romero-Rodríguez et al. 2018, p. 75–76).

Infoxication 2.0 reflects one possible answer to the producer's perspective of the million-dollar question: how do I make my product noticed in an immense sea of information? Being the first often seems to be the most important criteria, which increases the possibility of error and favors the headline instead of the background. Obviously, there is nothing wrong with breaking news. Anyone can make a mistake and later correct it—that is not fake news. However, the matrix of the web favors instant repetition—and that is what occurs immediately, mostly without confirmation. So, what happens when a user or a media outlet publish a lie? It will be repeated, sometimes like a snowball. Or when a number of bots disseminate the same information forging a large consensus? One piece of information may not be relevant if you follow 500 accounts on Twitter. How can you remember one specific fact in the midst of the fog? But the fact is that the repetition of a story increases exposition. It was already like that before the Web entered people's lives:

"This repetition may be central to audience retention of the information. When the media present the same or similar stories over a period of time, they are giving the audience a chance to mentally rehearse the information. Even within television news broadcasts, viewers are often teased with major headlines before stories are shown. Such rehearsal allows individuals to retain the information, even in cases of passive learning" (Tewksbury et al. 2001, p. 534).

That Is why Camus or Aeschylus gained so many adages on the web. It Is impossible that so many people are wrong.

The problem, one might even argue, is not the superficial true story, because if the topic is of any interest to the reader, he will search for some context. The problem is the weakening of the newsrooms that lead to an oversimplification of the narrative. The problem is when the reader is absolutely satisfied with an impoverished (but free of charges) version of what journalism should be. Background is the most important thing that journalism can provide. However, explaining the context requires time and people (like fact-checking does) in the newsroom, which is not the actual tendency. Some of my graduate students end their journalism internship without going out to the street with a senior reporter. Their work is desk only. And their biggest successes are measured by clicks and not by the actual impact of what they wrote. Reporting outside the newsroom or investigating seem to be endangered activities. Less people have to do more, but no one expects it to be better because there is no time. The case of the American newspapers offers a very clear example of this process. Newsrooms have cut down the staff to nearly 50% when compared to the period prior to 2008, despite "a modest increase in jobs after 2014 in other news-producing sectors—especially digital-native organizations— offset some of the losses at newspapers, helping to stabilize the overall number of US newsroom employees in the last five years" (Grieco 2020).

Drawing from our initial perspective, two other aspects are of paramount importance when looking at journalism today: the presence of disinformation in everyday life and the forces that circumscribe the freedom of the press and the Internet.

Let us begin by the global government censorship that is taking place all over the world (Curran 2018, p. 190). A number of cases, from Hungary to Serbia or China, show that we are far from having free media systems everywhere (Repucci 2019). Restrictions of freedom in the media are a recurring topic in democratic societies, mostly by their own cases of economic pressure or political interference. But the big picture seems to be absent from the public knowledge in those countries: "2019 was the 14th consecutive year of decline in global freedom," according to the Freedom House (Repucci 2020). The world is not getting any freer. People who think that this is a distant and exotic narrative are utterly wrong. Trump has proven otherwise. To be more precise, "more than half of the world's established democracies deteriorated" during this 14-year period and in 2019 only 42.6% of the countries analyzed are considered free (Repucci 2020).

In several democratic countries, a concerning number of nationalist movements has grown. Today, "where once democracies might have acted in unison to support positive outcomes to global crises, disparate authoritarian states now frequently step into the breach and attempt to impose their will" (Repucci 2020). This represents a concerning inversion of the democratization wave that followed the end of the Cold War, the fall of the Berlin Wall, and the collapse of USSR (Abramowitz 2019).

Reliable information is the main tool for fighting this trend. Having access to information that builds a strong public opinion has always been one of the fundamental premises of healthy democracies. In roughly two decades, the Internet scaled to the chief role in that process by becoming the main or one of the top source(s) of news in many countries (Shearer 2021; Newman et al. 2020). "Even in settings that are otherwise highly oppressive, an unrestricted online space offers immeasurable possibilities for free expression, community engagement, and economic development" (Shabaz and Funk 2020, p. 4). Information supports freedom of choice and self-governing. That is why restricting the freedom of the press or limiting the access to the Internet are part of the strategy of governments that sustain non-free regimes. Despite the limitations of their study, Kenski and Stroud suggested 15 years ago that the Internet could have a "positive and significant impact on political knowledge participation and efficacy" (2006, 187) which was considered a good sign despite not being the immediate solution for political disengagement. But that data was previous to the Web 2.0. On a more recent meta-analysis, however, the positive relation between the use of digital media and participation has revealed to be stronger in the era of social media (Boulianne 2018).

Internet use has also been associated with democratic diffusion in non-democratic countries (Placek 2020). Hence, the potential is clearly recognized. Nevertheless, the current situation is quite concerning. Internet freedom has also been deteriorating in the last decade, and the COVID-19 pandemic is accelerating this tendency, according to the Freedom House (Shabaz and Funk 2020). This organization has identified three main trends during 2020: "First, political leaders used the pandemic as a pretext to limit access to information. (...) Second, authorities cited COVID-19 to justify expanded surveillance powers and the deployment of new technologies that were once seen as too intrusive. (...) The third trend has been the transformation of a slow-motion "splintering" of the Internet into an all-out race toward "cyber sovereignty," with each government imposing its own Internet regulations in a manner that restricts the flow of information across national borders" (Shabaz and Funk 2020, pp. 1-2).

The last topic I want to emphasize is the fact that we live in a time of information disorder (Wardle and Derakshan 2017). We know that manipulation and fake news are an ancient phenomenon (Figueira and Santos 2019) on the media. The difference today is the immense quantity of information that flows through the web. Anyone can produce and distribute information online with no more than a simple smartphone. No special training is required to deliver a piece of video that ends up having the same exposition as a feature from the CNN.

The North American presidential elections of 2016 seem to be the pivotal moment that brought fake news to the attention of everyone (Allcott and Gentzcow 2017). Following the election, President Trump used fake news not only as a recurring argument but also as a strategy during its mandate. Fake news was adopted as a label for what was inconvenient, a political tool (Sullivan 2017; Levinson 2018). The expression was clearly instrumentalized depriving it from its meaning[5] (Sullivan 2017).

Nevertheless, fake news is only a part of the problem. That is why other expressions are better at framing what is going on. Claire Wardle and Hossein Derakshan congregate three notions to refer to the situation in a more comprehensive manner: "dis-information. Information that is false and deliberately created to harm a person, social group, organization or country. Mis-information. Information that is false, but not created with the intention of causing harm. Mal-information. Information that is based on reality, used to inflict harm on a person, organization or country" (2017, p. 20). The information ecosystem became more complex than ever. Information that is created by anyone spreads quite easily and is endorsed by legacy media or by a close friend on Facebook. People receive this immense quantity of information every day, a lot of which is fake, imprecise, or misleading. Moreover, the algorithmic matrix of the web environment can favor a biased reading of the facts and contexts [filter bubbles or echo chambers (Pariser 2011) are just an example] that is not always acknowledged by the users.

This is not the place to discuss and problematize the possibility of confirmation bias, conspiracy theories, naïve realism, bots, alternative facts, polarization, populism, or any other topic that is related to the use of information as a weapon of manipulation. The fact is that daily contact with manipulation attempts became a shrapnel from our relationship with technology. But most people are not ready to deal with that "info-war," as Aro (2016) calls it.

---

[5]Although it must be considered that fake news is an expression that refers to very different realities (Tandoc et al. 2018).

In the era of post-truth (McIntyre 2018) facts no longer constitute the congregating truth, which is terribly frightening, as we are realizing during the pandemic. "Post-truth is not about reality; it is about the way that humans react to reality. Once we are aware of our cognitive biases, we are in a better position to subvert them. If we want better news media outlets, we can support them. If someone lies to us, we can choose whether to believe him or her, and then challenge any falsehoods. It is our decision" (McIntyre 2018, p. 172). However, if want to know if Aeschylus said "In war, truth is the first casualty," and you find a thousand confirmations through Google, why would you continue looking? That is surely not the common user's behavior.

I agree with Umbelino when he says that we need a more substantial and structural change that embodies the spirit of the Enlightenment: "inform to fight the lie; inform to form a critical spirit; inform to emancipate; inform to question; inform to create the fascination of what is new and different" (2019, p. 176). That is why the author argues that worrying with improving the information has to be parallel and symmetrical to an investment in the development of the person (2019, p. 176).

People need to recognize and defend and support quality and reliable information as a public good. As part of a humanist view of the world, I believe that such an embracing and critical thinking will be able to impact the current spiral of decay that affects trust in democracy and journalism.

From the journalists' perspective, it is fundamental to be able to understand the *zeitgeist* and face these challenges knowing that this is not just about technology. A deterministic discussion will always be insufficient. It is more important to understand people's minds than what happens behind the screen, like Jenkins wrote: "convergence does not occur through media appliances—however sophisticated they may become. Convergence occurs within the brains of individual consumers" (Jenkins 2006, p. 3). This book offers an opportunity to deepen our knowledge on what is happening in the newsrooms and what may be the future of journalism.

January 2021   Sílvio Correia Santos
                                              Faculty of Arts and Humanities/CEIS20
                                                             University of Coimbra
                                                                Coimbra, Portugal

References

Abramowitz, M.: Freedom in the world: democracy in retreat. Freedom House (2019). https://freedomhouse.org/report/freedom-world/2019/democracy-retreat

Allcott, H., Gentzkow, M.: Social media and fake news in the 2016 election. J. Econ. Perspect. **31** (2), 211–236 (2017)

Aro, J.: The cyberspace war: propaganda and trolling as warfare tools. Eur. View. **15**(1), 121–132 (2016)

Barnett, S.: Will a crisis in journalism provoke a crisis in democracy? Polit. Q. **73**, 400–408 (2002)

Boulianne S.: Twenty years of digital media effects on civic and political participation. Commun. Res. 47(7), 947-966 (2020). https://doi.org/10.1177/0093650218808186

Bruns, A.: Produsage: towards a broader framework for user-led content creation. In: Shneiderman, B. (ed.) Proceedings of 6th ACM SIGCHI Conference on Creativity and Cognition 2007 (pp. 99–105). Association for Computing Machinery (2007)

Curran, J.: Triple crisis of journalism. Journalism 20(1), 190–193 (2018)

Figueira, J., Santos, S.: Percepción de las noticias falsas em universitarios de Portugal: análisis de su consumo y actitudes. El profesional de la información 28(3), e280315 (2019). https://doi.org/10.3145/epi.2019.may.15

Gaetani, G.: The noble art of misquoting Camus: From its origins to the internet era. J. Camus Studies 37–50 (2015)

Grieco, E.: U.S. newspapers have shed half of their newsroom employees since 2008. Pew Research Center (2020). https://www.pewresearch.org/fact-tank/2020/04/20/u-s-newsroom-employment-has-dropped-by-a-quarter-since-2008/

Jenkins, H.: Convergence Culture: Where Old and New Media Collide. New York University Press, New York (2006)

Kenski, K., Stroud, N.J.: Connections between internet use and political efficacy, knowledge, and participation. J. Broadcast. Electron. Media 50(2), 173–192 (2006). https://doi.org/10.1207/s15506878jobem5002_1

Levinson P.: Turning the tables: how Trump turned fake news from a weapon of deception to a weapon of mass destruction of legitimate news. In: Happer, C., Hoskins, A., Merrin, W. (eds.) Trump's Media War (pp. 33–46). Palgrave Macmillan, London (2019). https://doi.org/10.1007/978-3-319-94069-4_3

McIntyre, L.: Post-truth. MIT Press, Massachussets (2018)

Newman, N., Fletcher, R. Schultz, A., Andi, S., Nielsen, R K.: Digital News Report 2020. Reuters Institute (2020) https://reutersinstitute.politics.ox.ac.uk/sites/default/files/2020-06/DNR_2020_FINAL.pdf

Nielsen, R.-K.: The many crises of Western journalism: A comparative analysis of economic crises, professional crises, and crises of confidence. In: Alexander, J., Breese, E.-B., Luengo, M. (eds.) The Crisis of Journalism Reconsidered: Democratic Culture, Professional Codes, Digital Future (pp. 77–97). Cambridge University Press, Cambridge (2016)

Pariser, E.: The filter bubble: what the internet is hiding from you. Penguin Press, London (2011)

Pickard, V.: Can government support the press? Historicizing and internationalizing a policy approach to the journalism crisis. Commun. Revi. 14, 73–95 (2011)

Placek, M.: Learning democracy digitally? the internet and knowledge of democracy in non-democracies. Democratization 27(8), 1413–1435 (2020). https://doi.org/10.1080/13510347.2020.1795640

Repucci, S.: Freedom and the media 2019: a downward spiral. Freedom House (2019). https://freedomhouse.org/sites/default/files/2020-02/FINAL07162019_Freedom_And_The_Media_2019_Report.pdf

Repucci, S.: Freedom in the world 2020. A leaderless struggle for democracy. The Freedom House (2020). https://freedomhouse.org/report/freedom-world/2020/leaderless-struggle-democracy

Shabaz, A., Funk, A.: Freedom on the net 2020. Freedom House (2020). https://freedomhouse.org/sites/default/files/2020-10/10122020_FOTN2020_Complete_Report_FINAL.pdf

Shearer, E.: More than eight-in-ten Americans get news from digital devices. Pew Research Center (2021). https://www.pewresearch.org/fact-tank/2021/01/12/more-than-eight-in-ten-americans-get-news-from-digital-devices/

Sullivan, M.: It's time to retire the tainted term 'fake news'. Washington Post (2017). https://www.washingtonpost.com

Tandoc, E.C., Lim, Z.-W., Ling, R.: Defining 'fake news'. Digit. J. 6(2), 137–153 (2018). https://doi.org/10.1080/21670811.2017.1360143

Tewksbury, D., Weaver, A.J., Maddex, B.D.: Accidentally informed: news exposure on the world wide web. J. Mass Commun. Q. 78(3), 533–554 (2001)

Umbelino, L.: Para acabar de vez com as notícias. Elogio das pequenas coerências e da justa complexidade. In: Figueira, J., Santos, S. (eds.) As fake news e a nova ordem (des)informativa na era da pós-verdade (pp. 167–176). Coimbra: Imprensa da Universidade de Coimbra (2019)

Preface

There are times of paradigm breaking in journalism, either through the connections of readers' social networks, changing mass communication, or through mobility and ubiquity in the use of cell phones. We certainly live unexpected moments in social communication, for those who have been on this road for a long time as well as for new professionals in other areas, and students who think that it has always been so. Everyone in this business is surprised by the breakdown of communication paradigms; professionals who have migrated to journalism also witness the breakdown of their paradigms and those of communication organizations. Journalists, photographers, designers, and editors, everyone is being severely tested.

Unsurprisingly, Information and Communication Technologies (ICT) gave people the power to publish data, facts, events, and opinions. ICT also turned citizens into potential opinion leaders, making them, at the same time, consumers and producers ("prosumers") of information. The combination of "presumption" and the use of new products have allowed the emergence of new communication platforms that form an information ecosystem of its own and, unfortunately, do not prevent misinformation.

The breaking of the journalists' monopoly paradigm has been felt for some time. However, it is more dramatic nowadays due to the popularization of social networks and to a quasi-compulsion to use them. These are the times of new paradigms in social communication, disinformation, and post-truth. Such phenomena have led communication companies to seek survival, especially in the last decade. Sustainability has always been required from newspaper managers through different paths. Researchers are in charge of in-depth investigation and theoretical knowledge to help finding sustainable models. Studies identified the use of tools without the proper preparation of teams, and the adoption of new work procedures by increasing the mediation of ICT. In most cases, instead of facilitating management for executives not very fond of these practices, companies find it hard to set goals.

The chapters of this book present studies showing that traditional advertising business plans have changed as dramatically as the production of content. Advertisement in weekend newspapers comes from the twentieth century. In this ad segment, innovative companies with a global business changed the system. Ads are paid per view. The logic of pay per view has radically affected the traditional journalistic business models. In the context of the markets of journalism and advertising, these change led to more complex issues and new challenges to the survival of most journalistic organizations.

At the beginning of the current century, convergence processes were integrated into most newsrooms of local and national newspapers, groups of journalists, and publications for specific or specialized niches. Along with these ongoing processes, came the need to encourage collaboration in newsrooms, facilitated by information systems, and to review the workflow in the organization due to the presence of digital media, and also to choose data first for news production, and guide other activities from there. Nevertheless, before or in parallel, it was necessary to invest in computer and data journalism. The narrative should be centered in a large repository with the support of ontology or networked vocabulary (knowledge graph). It should provide greater accuracy and reliability of the information to face the increase in fake news and should increase the involvement with the communities and audiences that we intend to serve (Tsakarestau and Pogkas 2017).

The practice of collective editing of journalistic articles, distributed over time and space thanks to ICT, and the multichannel dissemination of news demonstrate that the factors mentioned are preponderant. They catalyzed the rupture of the centrality of newsrooms and transformed the production of information to virtual newsrooms, something contemplated in chapters of this book. These practices and context enabled the emergence of two concepts: (1) "deprofessionalization" of the main actor of newsrooms and (2) "deinstitutionalization" of journalistic organizations and transformation of the journalists' workplace.

We are experiencing new paths toward an interdisciplinary journalism. Mass communication was characterized mainly by the dramatic expansion of newspapers, radio, and television's reach. Technologies imply the use of new work methods, many times inverted. Journalism scholars continue to attempt representation through description or multi-faceted models, one of the issues this book focus on. In parallel, there is the arrival of the convergence of the media, the hybridity of the means or distribution channels, and the ubiquity allowed by cell phones.

The technologies bring instruments and procedures already incorporated in software developed for other service sectors. But they often present difficulties of use for older journalists, who reject them. The expectation is that information systems seek to be collaborative, and friendlier, and that artificial intelligence is embarked, probably in the algorithms, with tens of thousands of code lines. However, there is a real danger that intelligent agents would replace human beings if lawmakers do not establish ethical and moral limits, with universal adherence by nations around the world.

The search for newsrooms' models was the inspiration for a set of chapters. According to the theory of collective intelligence and the semantic sphere (Lévy 2013), the first requirement of the research was to be centered on the networked citizen and to facilitate the construction of creative and intercultural dialogues. The need to adapt content production came from the introduction of mobile devices and the presence of social media. Therefore, in order to survive in the hybrid society anticipated by García and García (2019), news organizations now operate in the media, and need to choose products and solutions that exploit as much as possible the capabilities of new devices, and social networks.

The expanded newsroom has become more sophisticated, also due to the spatial and temporal distribution, something that is well contemplated in this book. Now, the public space is also part of the newspaper, and everyone watches the rise of virtual newsrooms, a process also endured by Information and Communication Technologies (ICT) itself, when centralized CPDs shifted to distributed processing and then to cloud computing. In this scenario, newspaper professionals and communication specialists are increasingly dependent on ICT, which implies the need to include more technology in course programs of journalism. Students must know what is computational thinking and must have skills to use tools that allow handling the large volume of data available, in order to be able to deal with contemporary media (Pavlovic and Ljajic 2017).

Currently, some TIC tools facilitate the creation of transmedia content. In contrast, they reduce the difficulty in preparing and making available information more targeted to readers and sometimes help compromise the requirements of reliability and truthfulness. Some chapters of this book explore this aspect of the search for theoretical models in journalistic production. The new context should not reduce the professionals' social responsibility toward readers, viewers, and listeners. The new models for the production process must highlight the need for new functions for journalists and other professionals. Nevertheless, the opposite also happens, and many journalists need to better understand or accept the concepts and fundamentals, such as immersion, the capacity and limitations of each technology or solution (Hardee and McMahan 2017). In the same context, training is necessary for the use of tools, which are always replaced by new versions. Practices of innovation in routines and for the launch of new products are also needed.

The research project that gave rise to this book is entitled Multimodal Digital Media or MDM: a model proposal for a semantic framework of a collaborative environment for information management in a newsroom. The MDM project is detailed in the Appendix. The model should consider that the cooperative and distributed production routines would be managed from different ICT supports, with convergent digital media and a greater connection factor between sender and receiver, but within a humanistic and social context. Therefore, the construction of the semantic computational model of this research started from the understanding that the model could be composed of structured layers, something similar to network models, where a layer meets the request of the upper layer and requests services from the layer below it. The use of metalanguage based on the semantic approach of collective intelligence and digital media would establish the relationship between the elements of the model. However, it ended up arriving at a tetrahedral model with multiple faces, presented at the end of this book.

The model evolves throughout the project, and even after, to represent the process of producing journalistic articles. Among the justifications is the growing connection of people worldwide through social networks or digital media and the context of the mobility of communication. The first model used the concept of structured layers and then moved on to reliable figures such as the cube, where the cube's faces represent the essential functions of production. Finally, it evolves to the tetrahedron, where the management of flows, content, knowledge, and social networks are the main elements of the semantic and integrated computational model. However, it continued with the support of the Semantic Web tools. This means a convergent and integrated digital structure for contemporary newsrooms (Lemos 2018).

What is the book, given this explanation? It is the result of the work of the authors dedicated to a research project funded by the Brazilian Federal Government (CAPES/CNPq). They concentrated their efforts on the production of scientific articles and now on the construction of chapters related to research, which focuses on integrated and converging newsrooms and perspectives. The research project covered the fields of e-communication, computer science, and information science, all areas of the authors' expertise. The articles result from their technical visits to investigate newsrooms' productive routines and flows in major dailies from Brazil, Costa Rica, England, and the USA.

The prospects point to the production of midterm news. Factors are very similar to what happened in other service activities that had an increase in ICT-based tools, with an increasing connection from new media combined with the growing digital economy practices. Journalistic organizations have undergone digital transformations, and only those that have or have undergone full or accurate transformations must survive. In newsrooms, this includes the processes of searching, processing and distributing data and information, and the use of big data, with secure, automatic, and agile retrieval of information.

Objectives of the Book

The studies developed had their epistemological basis in the development of descriptive and exploratory research. They considered social and relational aspects to the work of communication professionals busy with such tasks in newsrooms. Other aspects were incorporated, such as learning to think and representing the accomplishment of tasks as a flow of information within the organization—as a workflow tool. We cite, for instance, the psychological effects of using interfaces with the computer, where many of their artifacts reside, be it systems, apps, and hitherto unconventional access (Karen 2005; Heravi and MCginnis 2015).

This book includes theoretical and conceptual research and empirical studies such as ethnographic research in newsrooms of major newspapers, application of methodologies in emerging issues of journalistic production, focus groups of professionals and researchers, and social network analysis in the research segments or areas of interest.

The first applications or experiments are already happening in the Communication Department of UnB, through the experimental publication *Campo Multimídia*, with theoretical support in the area of computer and information science. Other applications are under development.

The Chapters

The book consists of 10 chapters and comprises three segments. The first one presents preliminary results of the MDM project based on technical visits to selected news organizations. The theoretical basis and methodologies were established in the propositional documents and approved by the researchers. The second segment presents the consolidation of the methodology. The main scientific productions are achieved from the analysis of data and information collected and compared with other works available in the academic literature. The last segment of the book focuses on the evolution of the framework model for newsrooms and the development of systems for collaborative work environments, as well as the implementation of tools for news production and information centers.

The first chapter presents the results of a multiple case study, carried out in five newsrooms on three continents: *Correio Braziliense* and *O Globo* (both in Brazil), *La Nación* (Costa Rica), and BBC and Reuters (England). The analysis explores how technological convergence processes affect journalistic organizations and professional culture in accelerated production routines, reduced staff, declining revenue, and public participation. Some issues observed in newsrooms were explored, such as the insufficient use of new technologies impacting on people's activities. The first chapter also reveals several factors that shape the practice and the need to implement newsroom innovations (García et al. 2018). The qualitative analysis of the case studies reveals thoughtful and innovative perspectives on the examined issues; all reveal a journalistic profession undergoing deep and rapid changes as it adapts to a fast moving context.

The second chapter explores data journalism and professional profile in a comparative study in newsrooms across the USA, UK, and Brazil. It presents the phenomena of changes in journalism from different angles, some focused on cyclical changes, others on structural changes. However, the big challenging issue is to realize how much these changes—in professional routines and due to the adoption of technologies—effectively result in transformations. The chapter treats a critical problem in the context of modern journalistic practices and the journalist's identity—what is the job of a modern investigative journalist? Is s/he a programming buff, a mathematics expert, or a jack-of-all-trades? What role do data, big data, and data mining play in the modern journalist's identity? While the chapter does not provide definitive answers (and no exploratory study can), it does offer insights and perspectives from a variety of journalism stakeholders in reputed media outlets in Brazil, the US, and the UK. In the highly digitalized contemporary scenario, the chapter shows that the work of journalists has been significantly transformed and refined, due to a latent need to approach other areas of knowledge, specifically in the fields of computer science, information, and data science.

The focus of the third chapter is to analyze the workflow of the editorial staff at the Costa Rican *La Nación*. In the newsroom, actions of convergence and integration allowed the authors to understand the paper's modernization initiatives, thus prospecting the use of ICT in the treatment of content in newsrooms and news production. The study identified the non-functional and functional requirements in the newsroom's production information flow (workflow) as the basis for describing collaborative work environments and the use of IT. A digital transformation: The newsroom is still key to understanding contemporary journalism, but what would we see inside one located in a legacy media organization? One finding: a lot of empty chairs (Kosterich and Weber 2019).

The fourth chapter presents a case study that mapped part of the work environment in a newsroom, identifying the use of ICT and outlining prospects for technology in newsrooms. The daily *O Globo*, based in Rio de Janeiro, showed promising results for the future, and all the pros and cons in collaborative processes supported by IT. In order to understand the newsroom as a whole, the 4C collaboration model application is used with its dimensions of communication, coordination, cooperation and connection, and business process management tools to express workflow, allowing a preview of near-future news production. There are differences and changes between traditional employment patterns and positions in the newsroom, and the new requirements for journalists and professionals include new skills and knowledge of new technologies, involving programmers, coders, and data specialists (Kosterich and Weber 2019).

The fifth chapter seeks to identify the main functional requirements of the work environment in the journalistic newsroom, and collaborative tools and information systems in the dimensions of communication, coordination, cooperation, and connection. Moreover, we also wanted to identify software and frameworks used by collaborative technologies. We sought to verify how these IT tools support the management and production of journalistic content, with possible support of ontologies and Semantic Web standards. Based on a visit to the BBC's newsroom in Central London, the news production workflow in the newsroom was used as the main tool to analyze the processes of convergence. A major focus was on the workflow in the newsroom to understand and explore the process of convergence and integration.

The sixth chapter investigates the importance of the conversations and their potential to contribute to newsroom's production routines. This study was initially characterized as a descriptive, applied, and exploratory research based on the ontology of language. By developing the instrument of data analysis, called matrix of senses, the research was also considered methodological, based on grounded theory methodology (GTM). One of the essential contributions of this work is to situate the importance of the conversations, giving them a formal, theoretical, philosophical, and methodological visibility in the news production process.

The matrix of senses contributed to the explanation of patterns of behavior and presented itself as an instrument that can be customized and replicated to other contexts in which conversations play an important role. In the context of a conversation, journalists are expected to reskill, deskill, and upskill their practices and working routines, generally without any direct supervision. In doing so, they vulnerably move in and out of large and small newsrooms and news organizations, trying both to make a difference and to make ends meet in an exceedingly competitive market (Deuze and Witschge 2018).

The seventh chapter presents algorithms, numbers, and quantitative approaches that are changing Brazilian communication studies. It discusses the impact of transformations generated by the digital media ecosystem, from the quantitative explosion of emitters, sustained by the ubiquity of the binary networks and technological devices that support the production of content. By using design science and the digital methods approach, the authors explore the possibilities of epistemological and methodological expansion, based on interdisciplinary initiatives and the incorporation of new skills in the formation of professionals and researchers, in order to understand the current situation of data overload and tools poorly adapted. An application at LabCom/DCS/UFMA has identified four domains of knowledge that intersect to define the key requirements of immersive journalism: the fundamentals of immersion, standard immersive technologies, the fundamentals of journalism, and the main types of journalistic stories (Hardee and McMahan 2017).

This chapter addresses one of the core issues in contemporary journalistic production process: the increase in content and its storage in big data require the use of ICT and some expertise to do it. As the authors point out, there is a need for training journalists and communication professionals. It is no longer possible to outsource knowledge and skills in ICT. Professionals must know how to scrape the web, access API databases and services, use infographics in articles, learn computer language programming, and even the use of ontologies. This work has value for readers who may know the two more practical and useful research methods at the moment: design science and the digital methods approach.

In the eighth chapter, a reflection on origins, dissemination methods, contexts and confrontation of the disinformation culture, the authors report situations that promoted the distortion of facts at different moments in history, treating it as "disinformation" in the context of information science. A debate is proposed on initiatives and measures that can help raise people's awareness, improving their training for choosing and using the information in different situations. The article discusses possibilities that can contribute to curbing the spread of fake news and the culture of disinformation. To cope with the problem, the authors suggest the adoption, as a reference, of the concepts of digital literacy and information skills. Where are the actions to deal with disinformation? The basic answer is to use citizens' literacy horizontally, from children to adults. There are four relevant factors in the context of disinformation:

1. facilitated elaboration of content;
2. growing use of information by the digital economy forced by the pandemic;
3. the information war versus counter-information in globalization, nowadays also forced by the pandemic;

4. online newspapers with data journalism can be used to combat or generate fake news.

The ninth chapter brings the Semantic Web CMS to newsrooms. It proposes an ontology of viruses to improve the representation of knowledge in authoring environments. A semantic CMS is a tool that helps searching for texts related to the subject being written, while making semantic notes of what is produced. The challenge is to take semantic notes on a subject in the news details, since new subjects hardly have an ontology available on demand. This work describes the creation of the Zika virus' ontology by the project team, a new and relevant issue at the time of the epidemic, implementing annotation of news about Zika. The authors show the viability of using ontologies at the very moment of text production in a newsroom, unlike other approaches in which ontologies are used mainly for post-publication or to retrieve information

In the tenth and last chapters, the authors demonstrate how the consumers of news change their behavior of accessing and interacting with news content, of which they are now *prosumers* (combined news *producers* and news *consumers*). The challenge is to transform those proactive collaborators from a post-truth, era where political polarization and hate speech generate fake news, into a new renaissance period where a more supportive, generous, peaceful, tolerant, and inclusive environment produces reliable news in a virtual newsroom. Strict business processes and human analysis helped by AI tools could achieve it. In this scenario, applications such as WhatsApp, e-mail, Twitter, or Facebook are not only a source of information but also new channels of communication that publish customized content. Consequently, communication organizations face significant challenges posed by the decrease of paying readers and the competition imposed by emergent technologies that allow new ways to produce and disseminate news.

This book describes a unique moment in the newspapers' information production process. Firstly, the service sector and the media will have to invest in digital transformation, not only by investing in technology but also in its HR and the design of processes. The economy has moved very fast into the digital environment, resulting in a mass of digital illiterates. Therefore, the information will be a "commodity" when sponsored or directed to a specific objective, such as the cases of an epidemic or an advertising campaign. In the case of a specific and in-depth journalistic article, shaped for a recipient or a segment of society, for example, it requires delivering well treated information, and it might charge the reader. This means hiring talented professionals who use proper IT tools. On the other hand, general-purpose information must be carefully constructed, occasionally via robots, but ensuring that they do not mix with other unreliable sources, or even fake news, as this will be more and more frequent.

Brasilia, Brazil   Benedito Medeiros Neto
Coimbra, Portugal   Inês Amaral
Uxbridge, UK   George Ghinea
August 2020

References

Deuze, M., Witschge, T.: Beyond journalism: theorizing the transformation of journalism. Journalism **19**(2), 165–181 (2018)

García, J.A., Carvajal, M., Arias, F., De Lara, A.: Journalists' views on innovating in the newsroom: proposing a model of the diffusion of innovations in media outlets. J. Media Innovations **5**(1), (2018)

García, Á.A.V., García, X.L.: Diferencias y similitudes en la presentación de infografía en las apps y las versiones en línea de "El País" y "The New York Times". In: Narrativas jornalísticas para dispositivos móviles (pp. 279–301). Laboratório de Comunicação e Conteúdos Online (LabCom), UFMA, Brazil (2019)

Hardee, G.M., McMahan, R.P.: FIJI: a framework for the immersion-journalism intersection. Frontiers ICT **4**(21) (2017)

Kosterich, A., Weber, M.S.: Transformation of a Modern Newsroom Workforce: A case study of NYC journalist network histories from 2011 to 2015. J. Pract. **13**(4), 431–457 (2019)

Lemos, A.: Cibercultura e mobilidade: a era da conexão. Razón y Palabra **22**(1_100), 107–133 (2018)

Lévy, P.: The semantic sphere 1: Computation, cognition and information economy. John Wiley & Sons (2013)

Pavlovic, D., Ljajic, S.: Modern Media Technologies in the Education of Journalism Students. In: The International Scientific Conference, eLearning and Software for Education (Vol. 1, p. 211). "Carol I" National Defence University (2017)

Tsakarestau, B., Pogkas, D.: Data-led newsrooms: Integration, collaboration and workflow in data-first media organizations. For Future of Media and Communication Research: Media Ecology and Big Data International Conference. Fudan University, Shanghai, China (2017)

Zaragoza-Fuster, M.T., García-Avilés, J.A.: The role of innovation labs in advancing the relevance of Public Service Media: the cases of BBC News Labs and RTVE Lab. Commun. Soc. **33**(1), 45–61 (2020)

Acknowledgements

We especially thank Prof. Maria Emília Walter, Ph.D., Dean of Innovation and Research at the University of Brasília (UnB), for having believed and supported this project, giving institutional supervision and guidance, and being an inspiration in scientific research.

In writing this book, it was continually enhanced through our interaction and collaboration with many colleagues, to whom we respectfully and humbly offer thanks. Among them, Antonio Lisboa Carvalho de Miranda, Ana Elisa Almeida Ayres, Aurora Cuevas-Ceveró (MDM), Érica de Oliveira Carvalho (MDM), Fernando William Cruz (MDM/UnB Gama—FAG), Larissa de Jesus Silva, Luciano Pina Gois, Marcelo Bulhões Fonseca, Márcia Marques, Mônica Regina Peres, Nildo Moreira, Paulo Gustavo do Nascimento, Renon Pena de Sá (MDM), Sidney Ricardo Britto Villela de Medeiros (the cover), Thallita Silva, and Zanei Barcellos.

For their dedication and hard work, we heartfully thank all the authors and also the journalist Leda Beck, for her help on partially reviewing the English translation and formatting this document.

We would also like to thank newspapers companies for giving us an office space, ideas, and encouragement: Correio Braziliense/Brasília; O Globo/Rio de Janeiro; La Nación/Costa Rica; BBC London/England; Metrópoles/Brasília; and Reuters/London.

Finally, this book was only possible due to the support of the following institutions:

- Programa Ciência sem Fronteiras (CAPES and CNPq), Brazil
- Department of Communication (FAC), University of Brasília
- Department of Computer Science (CIC/IE), University of Brasília
- Department of Computer Science, Brunel University, UK
- Department of Information Science (FCI), University of Brasília
- Coimbra University, Portugal

- Rede Nacional de Pesquisa Aplicada em Jornalismo e tecnologia Digitais (JORTEC), Brazil
- Instituto Brasil de Tecnologia e Inovação (IBrTec), Brazil
- Associação Profissão Jornalista (APJor), Brazil

Contents

Editors and Contributors

About the Editors

Benedito Medeiros Neto is Professor at the Faculty of Communication of the University of Brasília. He received Ph.D. in Information Science at the University of Brasília. He is Researcher of the Computer Science Department of the University of Brasília focusing the Semantic Framework for Journalism. He is Associate Researcher at the School of the Future of the University of São Paulo in Digital Literacy and Mobile Learning. He is Reviewer at IGI Global, and also at the Ibero-American Magazine of Information Science, University of Brasília. He is Associate of the Journalist Profession Association. He participates at the following researcher groups of the National Council for Scientific and Technological Development: a) Applied Research Network in Journalism and Digital Technologies; b) Journalism and Memory in Communication; c) Technology and Digital Narratives); d) Competence in Information. He is Director of innovation and development at the Brazilian Technology & Innovation Institute. He was a Visiting Professor at Computer Science Department at Brunel University in 2018. His main research areas include Computer Science, Information and Communication; Network Engineering; ICT teaching; Informatics and Society; Collaborative Systems and Web; Semantic Web; Digital Inclusion; Digital Cities; Competence in Information, Social Networks and Evaluation of Innovation Programs.

Inês Amaral is Associate Professor and Director of the Undergraduate Program Studies in Journalism and Communication at the Faculty of Arts and Humanities of the University of Coimbra. She received Ph.D. in Communication Sciences at the University of Minho. She is Researcher at the Centre for Studies in Communication and Society and Associate Researcher at the Centre for Social Studies of the University of Coimbra. She is Coordinator of the Journalism and Society section of the Portuguese Association for Communication. Principal Investigator of the funded project "My Gender—Mediated young adults' practices: advancing gender justice in and across mobile apps." She is Member of the teams of the funded projects "SMaRT-EU," "(De)Coding Masculinities: Toward an enhanced understanding of media's role in shaping perceptions of masculinities in Portugal," "Opportunities and Challenges of Journalism in Open Environments." She is Consultant of the funded project "(De)Othering: Deconstructing Risk and Otherness." She is also Member of the Portuguese research team of the Global Media Monitoring Project 2020. She was Invited Scientist Fellow at Universidad Carlos III de Madrid within the founded project ENCAGE-CM, and Invited Professor at the University of Cape Verde. Her main research areas include audiences and media consumption; social networks, participation and social media; media and digital literacy; gender and media. She is a certified trainer by the Ministry of Education, the Union of Portuguese Journalists, and the National Scientific-Pedagogical Council of Continuous Education. She is Member of the Cyberjournalism Observatory and Media, Information and Literacy Observatory.

George Ghinea is Professor in Mulsemedia Computing in the Department of Computer Science, at Brunel University. His research activities lie at the confluence of computer science, media, and psychology. In particular, his work focuses on the area of perceptual multimedia quality and how one builds end-to-end communication systems incorporating user perceptual requirements. He has applied his expertise in areas such as eye-tracking, telemedicine, multi-modal interaction, and ubiquitous and mobile computing, leading a team of eight researchers in these areas. He has over 300 publications in his research field and is Editor in Chief of the *International Journal of Pervasive Computing and Communications*. Currently, his research pursuits are centered on extending the notion of multimedia with that of mulsemedia a term which he has put forward to denote multiple sensorial media, ie. media applications which engage three or more of the human sense. His work has been funded by both national and international funding agencies and has been covered by the BBC, Telegraph, and Forbes magazine, among others. He consults regularly for both public and private institutions in his areas of expertise.

Contributors

Inês Amaral University of Coimbra, Coimbra, Portugal

Suzana Guedes Cardoso Universidade de Brasília, Brasília, Brazil

Vitor Silva de Deus Universidade de Brasília, Brasília, Brazil

Maria de Fátima Ramos Brandão University of Brasilia, Brasilia, Brazil

Gentil José de Lucena Filho Conversational Intelligence Laboratory, Brasilia, Brazil

Thaïs de Mendonça Jorge Universidade de Brasília, Brasília, Brazil

Lillian Maria Araújo de Rezende Alvares University of Brasilia, Brasilia, Brazil

Ana Cristina Carneiro dos Santos University of Brasilia, Brasilia, Brazil

Márcio Carneiro dos Santos Federal University of Maranhão, São Luís, Maranhão, Brazil

Ébida Rosa dos Santos University of Brasilia, Brasilia, Brazil

George Ghinea Brunel University London, Uxbridge, UK

Gheorghita Ghinea Brunel University London, Uxbridge, UK

Lucas Hiroshi Horinouchi Universidade de Brasília, Brasília, Brazil

Edison Ishikawa Universidade de Brasília, Brasília, Brazil

Benedito Medeiros Neto University of Brasília, Brasília, Brazil

Edgard Costa Oliveira Universidade de Brasília, Brasília, Brazil

Gislane Pereira Santana Universidade de Brasília, Brasilia, Brazil

Elmira Luzia Melo Soares Simeão Universidade de Brasília, Brasilia, Brazil

List of Figures

List of Tables

Journalistic Newsrooms: Convergence and Innovation on Three Continents—A Case Study on Five Media Organizations

Thaïs de Mendonça Jorge and Benedito Medeiros Neto

Abstract This article presents the results of a multiple case study carried out in five newsrooms on three continents: *Correio Braziliense*, *O Globo* (Brazil), *La Nación* (Costa Rica), and BBC and Reuters (United Kingdom). The purpose of the analysis was to explore the effect of technological convergence on journalistic organizations and professional culture at a time when production routines are accelerating, staffs are being reduced, revenues are declining and public participation is increasing. The methodology consisted of visiting the newsrooms and conducting interviews with journalists. The analysis results confirm that there are doubts and uncertainties about using new platforms—which includes social media—and that there is a kind of experimentalism when it comes to using advanced resources despite the common ground that media organizations share: innovations are disseminated and the newspaper industry is forced to adapt.

Keywords Newsrooms · Convergence · Innovation · Newspaper · Media organizations

1 Introduction

Convergence is a buzzword these days. In the field of communication, it is hard to find a scientific article in which the word does not appear. Even though not directly referring to journalism, Iosifidis (2013, p. 172) defines convergence as "the delivery of media, telephony and Internet services through the same transmission platform, whether similar, existing or new." The author analyzes the issue at three levels:

T. de Mendonça Jorge (✉) · B. Medeiros Neto
Universidade de Brasília, Campus Darcy Ribeiro, Brasília 70910-000, Brazil
e-mail: thais.mendonca@fac.unb.b

B. Medeiros Neto
e-mail: medeirosneto@unb.br

(a) technological—the digitalization of broadcasting networks, information technology (IT), and telecommunications;
(b) structural—corporate partnerships in different sectors; and
(c) services and markets—added value and multimedia services.

Iosifidis (2013, p. 172) understands that technological convergence is a reference mainly to the integration of services: "Newspapers can be read online. Cell phones are now compatible with television." This is possible because the development of interactive digital broadcast services has led to a convergence between telecommunications and broadcast media. The rate of convergence is influenced by a series of structural changes in the information and communication industries, such as mergers and acquisitions. Corporate consolidation means that traditional entities seek new businesses, opportunities, and streams of revenue, using horizontal and vertical integration (Iosifidis 2013; Deuze 2004) as an innovation tool. Horizontal integration is the consolidation of media organizations that are involved in the same part of the production process; e.g. content production, and its distribution across platforms. In contrast, vertical integration is characterized by companies involved in different parts of the production process (e.g. distribution and marketing) or a newspaper associated with a bank.

In a generic sense, media convergence is the fusion of mass communication markets (print, television, radio, internet) using portable and interactive technologies on digital presentation platforms. Kawamoto (2003, p. 4) lists hypertextuality, interactivity, non-linearity, multimedia, and customization as some of the characteristics of digital journalism; in other words, convergence. He defines it as "the mixture or union of historically separate technologies and services". Palacios shares a similar line of thinking in his definition of online journalism, describing it as having the properties of interactivity, hypertextuality, personalization, memory, and instantaneous access, as well as multimedia/convergence. Multimediality/convergence "refers to the convergence of traditional media formats (image, text and sound) in the narration of the journalistic fact". Gradim (n/d) reduces convergence in journalism to four distinct realities: the convergence of economic groups, *media*, and newsrooms; how to collect and present news; and the multimedia product, which is available to the public.

This article presents the preliminary results from a multiple case study performed on five newsrooms on three continents: *Correio Braziliense*, *O Globo* (Brazil), *La Nación* (Costa Rica), the BBC and Reuters (England). This work is part of a project entitled "Multimodal digital media in newsrooms: a semantic computational model in a convergent digital structure. The study of information systems in Brazil, Costa Rica, England and the United States.[1]"

---

[1]The study used resources from Edites Capes 09/2014 Special Visitor Researcher. Under number 88881.068354/2014-01, it lasted 36 months and, in its multi, inter and transdisciplinary sense, involved the areas of Communication and Computer Science at the University of Brasilia and Brunel University in London, with the participation of British professor George Ghinea.

The objective of the analysis was to explore how technological convergence processes affect journalistic organizations and professional culture, in a scenario of accelerated production routines, reduced staff, declining profitability and increasing public participation. Visiting the newsrooms in South America, Central America and Europe showed us how media organizations are all subject to the same globalization and digitalization influences; however, they are affected differently by the insertion of technologies, and thus seek their own specific solutions while considering innovation and modernization of structures.

2 The Evolution of Convergence

The media convergence movement developed from recent technological advances, mainly the emergence of the Internet and the digitization of information. Deuze (2004, p. 143) highlights that convergent multimedia structures have been emerging since the mid-1990s, with companies from all over the world choosing some form of cooperation or synergy between their employees, newsrooms and departments, areas that had previously operated individually. This author also points out that some journalists think that for the media industry it is primarily a cost saving manoeuver.

The idea of converge is to approach, join or evolve in a positive sense; to take advantage of similarities shared between organs and use those similarities to develop new organisms, introducing a kind of change or mutation that occurs on different levels. The convergence phenomenon presupposes technological, market, cultural and social changes (Jenkins 2006). Bringing convergence to the journalistic sphere, according to García Avilés et al. (2009, pp. 173–198), is a "process (…) that affects the technological, business, professional and editorial side of media".

In short, there are four dimensions of convergence related to the news industry: (1) technological convergence—the more obvious aspect, it goes beyond network transmission to combine content, languages, teams and support; (2) business—this occurs in mergers, incorporations and diversification of activities; (3) professional —this focuses on the productive structures and often reshapes and constrains the newsroom environment; (4) content convergence—influences *multiplatform* or *multimedia journalism* (Fernandes and Jorge 2017).

Salaverría and Negredo (2008) draw attention to the fact that the integration of newsrooms in journalism is the most tangible element of the convergence process, while also being the most complex. It goes beyond just job restructuring and staff reductions as it focuses on routines and journalism. Digital technologies are used in journalism to ascertain, produce and make news available "to an audience which is becoming increasingly familiar with computers" (Kawamoto 2003, p. 4). This is only one of the elements of journalistic convergence as it pertains to the product itself, which generates other types of fact narratives.

The characteristics and peculiarities of each element, including its reach and target audience, make developing a unique formula or recipe for convergence difficult. When newspaper companies began digitizing their newsrooms (by replacing typewriters with computers, preceded by a rationalization of graphic processes) they were taking the first (unconscious) step towards the technological convergence we see today. Larrondo et al. (2016, p. 280) describe the reason to the ascent of convergence by "the need to provide a better news service for all audiences, both offline and online", but see budget restrictions as well.

These developments in convergence included changes to how technological facilities were used, to the organization of the publication and distribution processes, to the restructuring of the physical workplace, in addition to changing the framework of newsrooms by creating or eliminating positions and installing a new culture in news media. Convergence has already been viewed as a product (during the 1990s), as a system (2000), and has only recently come to be viewed as a process (or various processes), one which requires studying it at a higher level of complexity, where organizations, people and technologies are analyzed in the same spaces of interaction but subject to different flows, taking sociological, local, strategic and environmental aspects into account.

Appelgren (2004, pp. 237–248) observes the convergence in each stage of news production: the creation, edition, distribution and consumption of information. She goes on to state that we must not confuse the convergence process with its effects, "since it is the outcome of the processes that affects us, not the process itself". The effects might be visible, detectable, while the process not. As we will see, the process of convergence can be strategically planned; it is influenced by market forces, and subjected to tendencies in society, as well as to economic and social sceneries, and technological developments. For the media industry, these effects could be:

(a) integration;
(b) combination;
(c) merger;
(d) cooperation; and
(e) cross-promotion.

Salaverría and Negredo (2008, pp. 177–181) list the five phases of convergence in the news industry:

1. the digitalization of newsrooms;
2. the installation of newsroom structures through online journalism;
3. the physical integration of traditional and online newsrooms;
4. the development of new languages according to the public's interest;
5. and the fusion of structures, making them indistinct.

What we are talking about in this article is the technological and structural convergence understood by Iosifidis (2013) as horizontal convergence, in which, media organizations are located in the same part of the production process, the production of content and the distribution occur by various platforms which, as we know, is circular in nature and affects the professional culture and the content produced.

3 Methodology

The research methodology used in this study was composed of a qualitative, exploratory, and descriptive approach to explain the different points of view, in conjunction with the case study, in a real context of journalistic newsrooms. Certain limitations to the field survey (such as time) made it impossible to study all the variables related to the object of study. Data collection was obtained through semi-structured interviews, *in loco*, conducted with specialists, managers and journalists. The purpose of the interviews was to know their perceptions of various elements of this ongoing convergence. The organizations selected for this study (market leaders in their segments) were willing to narrate their own processes. Other organizations included as part of our initial idea refused to participate.[2]

We initially considered doing an ethnographic research in the newsrooms of the various countries. This proved to be unfeasible: having researchers conduct interviews within the journalistic environment has been increasingly difficult due to editorial decisions and to each individual media vehicle's strategy. Nevertheless, *O Globo*, *Correio Braziliense* and *La Nación* were very accommodating and allowed their journalists to be interviewed in the workplace. Researchers conducted interviews in *O Globo* and *La Nación* for three subsequent days and in *Correio Braziliense* on two separate occasions. Visits were limited to 2 h in Reuters and 5 h and 15 min in the BBC. In those cases, researchers were not given open access to journalists and the sources of information were limited to those who received the group from UnB/Brunel.

On previous occasions, when we performed a kind of informal pre-test in the newsrooms, we noticed that the word "convergence" was not used, and the meaning of this term as it is used in theoretical studies was not widely known. We asked all managers/journalists who participated in the study the following question: "What does convergence mean to you?" This is important because there could be differences between what theorists think convergence is and how convergence actually occurs (or not) in practice. We therefore needed to know the opinions of those who were directly involved in implementing newsroom processes.

4 Results

The results not only show parallels but also differences in the styles and practices of internal organizations. The following results demonstrate different aspects of challenging converging processes identified in five newsrooms on three continents, each one having their own culture, news consumption habits, and production and distribution routines.

---

[2]*The New York Times* has repeatedly refused to participate; *The Guardian* requested remuneration, which the project did not have the budget for.

4.1 Correio Braziliense

Correio Braziliense—the main newspaper in the capital of Brazil—first went online in August 1996 with *CorreioWeb*. Up until April 2008, *CorreioWeb* was the official website for the *Correio Braziliense* newspaper. The site was redeveloped that same year (2008) with the aim of "making it more interactive". It was divided into two digital companies with different legal entities: correioweb.com.br and correio-braziliense.com.br. The former maintained its independence and to this day continues to inform on civil service examinations and entertainment. The latter focused more on news, making it more of a national vehicle. This newspaper tried to model itself after *The Washington Post* (*U.S.A.*), a newspaper in the US capital.

Despite taking steps towards structural convergence and operating in a joint newsroom (paper/online), the correiobraziliense.com.br portal, which belongs to the Diários Associados Group, has always suffered from severe budget restrictions and could be an example of why it is not enough for structures to just simply be joined together. Since 2010, the biggest newsroom in Brasilia have depended on interns,[3] who come from a number of journalism courses in the city and take on the role of a journalist, including the responsibility that comes with being a true professional. If some authors point out that convergence is a collaborative and cooperative way for newsrooms, which used to operate separately, to work together (Deuze 2004; Boczkowski 2003), we found that the online and paper teams at *Correio Braziliense* have been unsuccessful at integrating with each other (Fernandes and Jorge 2017).

There is one team which feeds the site 17 h a day, and the articles it produces are often used for the printed newspaper; however, it is not common for print journalists to contribute online, and when they do, it is only under request. Visiting the *Correio* (an average daily circulation of 53 thousand, including both printed and digital signatures (IVC 2019; *apud*, Rosa 2019) allowed us to see that journalists from both platforms share the same physical space, but there seems to be an insurmountable barrier between them. The reasons for this barrier seem to result from the lack of an integration policy, a lack of education and training for these professionals, not to mention feelings of prejudice and rivalry among young people who feed the digital spaces. Perhaps this is why nobody talks about convergence. When we ask about issues related to the systems used, we are addressed to the younger team of the IT sector who supports the newsroom; and when we interrogate about the *modus operandi* of digital and print journalists, they indicate the "integrated" newsroom (Executive Editor, personal interview, July 3, 2015).[4]

---

[3]Comprised of a team of 45 professionals from Correiobraziliense.com.br: 28 trained journalists, 10 interns, and seven technicians. The information was collected from the interviews and from the visit to the newsroom in August 2010.

[4]For ethical reasons, we opted not to mention the real names of those interviewed.

4.2   O Globo

With an average daily circulation of 315 thousand copies, *O Globo* competes with *Folha de S. Paulo* and *Super Notícia* for the title of the most widely circulated newspaper in Brazil.[5] In the *O Globo* model, journalists work for the news *core* regardless of the destination of the content (paper, website, or other media group vehicles. Like many communication companies, Group Globo launched popular versions of its newspapers, such as *Extra*, which currently coexists under the same roof in a new construction.[6]

According to the Editorial Director at *O Globo* (personal interview, September 8, 2015), the objectives behind the first actions of convergence were to better explore the market, to seek a new business model, to face competition, to launch new products in market segments, and to improve future prospects. Convergence has taken place in several stages over a long period of time, and continues to do so. In the meantime, new work models have been looked at with efforts being made to improve the flow of information. In the meantime, workplaces were being reshaped through the construction of a new building and the idea of a "multimedia newsroom".

For the Executive Multimedia Editor at O Globo (personal interview, September 8, 2015), the culture of print newspaper professionals was a barrier preventing the team from achieving a higher performance. The transformations that convergence brought to the newsroom (referred to as "integration") were met by a natural resistance to ICT, and a cultural distance, especially from more experienced journalists. The executive editor stated that all the changes were made in order to adapt the workflow of the newsroom to that of a multifaceted newspaper with content

---

[5]At the end of 2018, *Folha de S. Paulo* newspaper was second in terms of total circulation with 310,000 printed copies and digital signatures. In third place was *O Estado de S. Paulo* with 239,000 copies, and in fourth place was the popular newspaper *Super Notícias*, from Belo Horizonte, with 184,000 copies. The sixth place was *Correio Braziliense* (53,000 copies in total circulation). This data was provided by the IVC, cited by *O Globo* (Rosa, Bruno. "*O Globo* is the newspaper that grew the most in 2018", Jan. 25, 2019. Available at: https://oglobo.globo.com/economia/o-globo-o-jornal-que-mais-grew-em-2018-23400125 (access on Jan. 29, 2020). The decline in print newspapers continued in 2019. The digital vehicle *Poder 360* reported, according to data from the IVC, that the daily print circulation for the major Brazilian newspapers in October 2019 was as follows: *Super Notícias*, 140,387; *O Globo*, 104,129; *O Estado de S. Paulo*, 97,125; *Folha de S. Paulo*, 86,196; *Correio Braziliense*, 16,409. Since 2014, the drop in circulation reached 71% in some cases, such as the *Estado de Minas*. (*Poder 360*, "Newspapers in Brazil lose circulation and digital sales remain modest," Nov. 26, 2019. Available at: https://www.poder360.com.br/midia/jornais-no-brasil-perdem-tiragem-impressa-e-sale-digital-still-and-modest/. Access on: Jan. 29, 2020).

[6]We arrived at *O Globo* while it was still located at its old headquarters on Rua Irineu Marinho, no. 35, on September 8, 2015, the day after the Brazilian Independence Day holiday. The editorial staff was experiencing massive layoffs: about 40 employees (mostly journalists) had been dismissed, reaching a total of 100 employees laid off recently. As for the newsroom, it was reduced to just a little over 200 journalists. The company claimed that the reason for said reduction was to "adapt to the business model."

distribution in line with the twenty-first century reader. "The newsroom focuses its attention on production, which means the end of the print newspaper, a major reference point for *O Globo*'s work and its main source of income. The print newspaper continues to be important, but we are preparing to provide qualified information across all platforms", said the executive, who came to the vehicle after working for the Spanish consultancy Cases i Asociates, the agency responsible for one of the reforms that *O Globo* undertook toward digitalization.

Although interviewees at Globo agreed that new newsroom management models needed to be implemented, including the addition of new hardware and software, in practice, its adoption was not so simple and immediate. If we look at the pros and cons, we cannot say for sure that the work of journalists has become less important than the method or model being implemented. However, it is possible to predict a drop in the quality of news content if there is any kind of forced or unplanned process. In this case, to keep up with their vision, the leader of the convergence process could make their point of view more widespread and thus maintain the pace of integration at the expense of accepting new rules, which would have consequences for the product and perhaps even for the credibility of the brand.

The introduction of more flexible working hours (for individuals and for teams), the convergence between print and digital, and the greater use of collaborative systems to facilitate integration continue to present challenges to the *O Globo* newsroom. One cannot disregard the conflict between generations, something that had hoped to be resolved with new information flows and the formation of mixed teams composed of more experienced professionals and young, recent digital graduates (Editorial Director, personal interview, September 8, 2015) who have more affinity for new technologies.

Even though the current business model still prioritizes print newspapers due to the financial return they represent, the concentration of media and professional support such as infographics, design, photography, art and layout, and the centralization of the complex Infoglobo[7] operations in a new building are geared toward rationalization and economics. Journalists still remain divided, which makes cooperative work more difficult and increases hierarchy, as a result, proactivity works best with younger journalists (Editor 1, personal interview, September 8, 2015).

4.3   Reuters

News agencies are newspaper companies that specialize in distributing data and news from the sources of the event to media outlets such as newspapers, magazines, radio stations, websites, and television stations. The first agencies appeared in the

---

[7]Infoglobo is the corporate name that encompass the products *O Globo*, *Extra* and *Expresso*, the Globo and Extra websites and Agência o Globo.

mid-nineteenth century, which included the Havas news agency (now France Presse), founded by writer and translator Charles-Louis Havas in 1835.

Two Havas employees, Paul Reuter and Bernhard Wolff, went on to set up two agencies (in London and Berlin, respectively) that would compete with Havas: Reuters in 1851 and the Wolffs Telegraphisches Bureau in 1849. The Reuters story got its start in Aachen, on the German border with Belgium and the Netherlands, where homing pigeons were used in 1850 in order to help transmit financial news (Reuters 2019).

Reuters, the news and media division of the Thomson Reuters corporation, is currently the world's largest international multimedia news provider and was named the number one international brand for digital reach in an Ipsos Affluent Europe 2017 survey, placing it ahead of 44 other international media brands. It reaches more than a billion people every day. Reuters provides business, financial, national and international news to professional media market clients around the world.

The Reuters newsroom (and headquarters) in London is the largest in Europe and is located in the economic heart of the city, in Canary Wharf. At the time of this study, it was a traditional space that had been adapted to the technological era. A large TV/radio studio was added to the journalists' room, which customers could rent to use for exclusive reports and interviews. The newsroom was divided into three main areas: text, TV and Internet. Text (including photography) still remains the biggest area and followed the agency's tradition of distributing exclusive content from its correspondents (reporters and photographers) around the world.

The text area maintains an editorial structure and, although Reuters does not specialize in opinion-based reporting, it instead provides basic and "bias free" material (Reuters Institute 2014), the subscribers ended up forcing the agency to create a sector for analyses and opinions, which today operates in the newsroom environment. However, at first glance this place looks chaotic and archaic due to the large number of people in small, individual and sequential tables, and the mountains of papers. This old scenario seems to bother journalists, who look forward to changes. In a few months, the Reuters' London newsroom would be modernized and gain a new space in the same building (Assistant Executive Director, personal interview, May 29, 2017).

The idea of a convergent newsroom was not foreign to the Executive Director, who was enthusiastic about the alterations that should make the environment more organic and rational, including a central desk. As an international news agency, Reuters encourages and strives for a tradition of cooperative working that helps maintain a dedicated team of proofreaders, a position already eliminated in most other vehicles. Reuters has 14,000 employees; 2500 of them are journalists operating in 204 cities and broadcasting news in 19 languages.

4.4   The BBC

When looking at newsroom models prevalent in the Western world—the *open space* model as used in the film "*All the President's Men*" or the more recent movie "*Spotlight*"—the BBC's in London is different, mainly because of the mobility of teams in the workspace. The BBC is a public TV broadcaster formed in 1922 and funded by United Kingdom taxpayers who pay a monthly fee to enjoy ad-free services 24 h a day. It transmits TV, radio and Internet broadcasts nationally and abroad, and has almost 19,000 employees with an operating cost of GBP 5 billion. Its headquarters is the Broadcasting House located in central London. It is a modern newsroom that has no fixed places.

"People move from table to table," says BBC News Senior Product Manager (personal interview, May 17, 2016), explaining that when an important and immediate issue breaks, journalists and other professionals work together. There is also a central decision table where most of the decisions are made; however, certain TV programs do have their own structure. A series of giant-sized screens—similar to the ones used at airports—are set up so that the information arriving from the field is made available to everyone; in this way, those responsible for each sector can screen the information and forward it to specific channels.

The BBC has four official editions:

1. the United Kingdom;
2. the United States;
3. Asia; and
4. the rest of the world.

There is also a special section dedicated to User Generated Content (UGC). For example, the newsroom checks a tweet and quickly places it on web pages that are immediately generated by the teams working together. The BBC recognizes that photos, videos and audio sent in by users for any kind of unexpected event helps streamline coverage and adds value to the news (Product Manager, personal interview, May 17, 2016).

4.5   La Nación

Costa Rica is a prosperous country with a more diversified economy than the rest of Central America, and has adopted strong neoliberal policies. *La Nación* Group publishes the largest circulation newspaper in Costa Rica, with an average of 100,000 copies per day. The process of integrating media vehicles (the newspapers *La Nación*, *El Financiero*, *La Teja*, and other monthly magazines) started in 2007 and ended in 2011 with the relocation to a building constructed specifically for this purpose. It is the only building in Central America with an integrated newsroom. It cost US\$ 7 million. The newspaper has about 440 professionals, made up of 270 journalists and 170 professionals from other areas (IT, design, specialists, etc.).[8]

The newspaper's Strategic Plan is its main guiding tool, headed by general management and another task group of seven journalists who receive help from IT and Production Engineering specialists. The entire newsroom, which was reinvented through the help of Spanish firm *Innovation Media Consulting Group*, was committed to changing the culture of traditional print media and to start thinking about content that was previously available on the website, on social networks, smartphones and tablets. This strategy also led to changes in its business model, such as a more aggressive approach to increasing digital subscribers, a strategy that *O Globo* has also adopted.

In this sense, the journalists at *La Nación* were involved in the concept of *digital first*; a term used to refer to actions that should be directed to the digital area first. At the time of our visit (2016), the Sports section had managed to completely incorporate this concept. The newsroom professionals, who complained about the model that tried to be imposed, said they were committed at that time to "tropicalize" the ideas (Executive Editor, personal interview, January 25, 2016) as they considered some initiatives to be "too European"; such as establishing fixed closing times for each section. However, the new structure of the newsroom, the *paperless* principle, and even the inclusion of new sections (such as Radar and Eco) were adopted with relative ease.[9]

5 Discussion and Conclusions

Convergence in journalism represents an opportunity for print media, audiovisual media and the Internet to cooperate and produce complex news; i.e., using digital resources to give users the most complete information possible. Convergence is not an obligation in the market; however, it can be used to avoid the crisis in the journalistic sector while also reducing physical functions and structures within a company and offering new products through innovative narrative solutions and distribution systems capable of reaching a wider audience.

Convergence is a developmental process involving various social actors—entrepreneurs, journalists, sources, and the public—and the effects of these changes are open. We are currently living in a time when technological convergence is all around us (on cell phones, at the bank, at home, on the street, at work), and for newspaper companies, it has been brought on as a consequence of globalization which raises concerns about capitalization and competition. It is by no means a linear and peaceful cycle from the point of view of the information industry, and

---

[8]Personal interview with the CEO, January 25, 2016.

[9]Journalists stated that *La Nación* had made radical cutbacks to staff in the previous year, reducing the newsroom in half. Those professionals who did not adapt to the new model left.

even less so from an employee's perspective. Like any change, this one is also sudden, although it necessitates a long and complicated preparation process, one involving intense symbolic exchanges and transformation in the culture of journalists.

Studying the examples in this paper led us to the conclusion that convergence is a maturing process. There are no recipes to follow. Companies are experimenting with digital tools and, driven by competition and haste, they are looking for their own models to circumvent problems without reducing the quality of the news product and also to serve an increasingly conscious and demanding reader. Slowly but surely, the media industry is leaving the old division between media and vehicles behind, and in its place is assuming a convergent profile: a multiplatform complex with multitasking journalists capable of producing customized information. However, this does not come without a cost.

We also saw that this is not a painless process: it leads to precarious working conditions for journalists, a loss of referents in professional culture (from the old *legman* journalist who reports from the scene of an event, to the journalist who reports from behind a desk), and changes to ideologies, roles and social functions. Not to mention changes to the logic of the news product which was previously seen as a symbolic asset but is now viewed as a *commodity*.

According to the five phases of convergence from Negredo and Salaverría (2008), we can say that all vehicles studied have reached the first phase: the digitization of newsrooms. Although the resources and methods they have adopted may differ, they all create their own editorial structures to facilitate online journalism (Phase 2). Phase 3 (the physical integration of traditional and online newsrooms) occurs with restrictions and certain unease among workers, but those who have experienced it for longer periods of time show us that it tends to be natural. Phases 4 (new languages) and 5 (fusion), in what would be total convergence, are under development in more mature structures such as those of Reuters and the BBC.

In Brazil, the more economically viable media organization, such as *O Globo*, have been more successful at assimilating innovations in the newsroom. *Correio Braziliense* takes two steps forward but then one step back as it sometimes cannot obtain enough support or financial resources, or does not decide on or face the scenario of changes. As Soria (2011) states, "true integration needs people who think about the news in terms of a driving force, like a turbine in an industry". He also believes the idea of saving by reducing structures is wrong: "Integration requires good professionals who are well paid and dedicated. Integration for the purpose of saving is not doing quality journalism" (Soria, personal interview, February 6, 2020).

The fact is that all newspapers in South and Central America have been looking for viable business models and to adopt efficient *workflows* for their newsrooms as revenue from digital products and services is not enough to support the operating costs of the newspaper company, which currently operates on a ratio of 1:10. Even with the facilities that new technologies have provided (which are becoming more accessible day by day), producing content for data journalism, text journalism, and audio and video journalism is still expensive. The content market is in decline and advertising continues to drop (Bertocchi 2013). The most advantageous business model for digital advertising is still with OTT (Over-The-Top media services such as Facebook, Google, Netflix, You Tube). These Internet companies, which quickly deliver media to specific target audiences, use state of art technologies and thus have an advantage over traditional news companies.

On the other hand, horizontal integration (Appelgren 2004; Iosifidis 2013; Deuze 2004)—the consolidation of media organizations with the production process—has occurred with all the subjects analyzed in our sample through content production and multiplatform distribution, even in different formats. Vertical integration—companies associated with other companies from different sectors of the economy—was only verified in the case of Thomson Reuters. From the results of the convergence process highlighted by Appelgren (2004), we could see, in our study, the effects of integration, combination, and merger (of companies or the organizations within them), and the cooperation between teams; however, we did not detect any "cross promotions", defined as cooperation and promotion among partners through their media channels.[10]

We are aware that convergence is not a term used by all journalists; many of them prefer to call the process integration. One of the interviewees even had to ask us to clarify what we meant by convergence. Convergence remains a very generic concept even among reporters and IT staff. They seem more concerned with how the decisions of management will affect their daily lives, with maintaining their employability and the conditions they need to perform their jobs.

One point that could be taken up in further studies could be how the organizational culture deals with the overload of work in the face of layoffs and the lack of appropriate wage compensation, in addition to problems related to the technology itself, to more experienced journalists accepting that technology and the objectives intended by managers, owners or shareholders of newspapers.

References

Appelgren, E.: Convergence and divergence in media: different perspectives. In: 8th International Conference on Electronic Publishing, June 23–26, Brasília, pp. 237–248 (2004). Retrieved from: https://www.researchgate.net/publication/37676508_Convergence_and_divergence_in_media_Different_perspectives

Bertocchi, D.: Dos dados aos formatos—Um modelo teórico para o design do sistema narrativo no jornalismo digital. Doctoral dissertation presented at the School of Communications and Arts, University of São Paulo (ECA/USP) (2013)

---

[10]Please see also Dailey et al. (2003): Dailey L., Demo L., and Spillman M. The Convergence Continuum: A Model for Studying Collaboration Between Media Newsrooms. Ball State University, Muncie, Indiana, USA: Paper submitted to the Newspaper Division of the Association for Education in Journalism and Mass Communication, Kansas City, Missouri, July–August. Maybe in a study on the content shared by *O Globo*, *Extra* and *G1*, for example, we could notice those effects of cross-promotion. But this is not a focus of our research.

Boczkowski, P.: The construction of online newspapers: patterns of multimedia and interactive communication in three online newsrooms. Paper presented to the ICA Communication Borderlands Conference, San Diego, CA (2003, May 23–27)

Deuze, M.: What is multimedia journalism? Journalism Stud **5**(2), 139–152 (2004)

Fernandes, S.G., Jorge, T.M.: Routines in web journalism: multitasking and time pressure on web journalists. Braz Journalism Res **1**(1), 20–37 (2017). https://doi.org/10.25200/BJR.v13n1. 2017.909

García Avilés, J.A., Prieto, M.C., Kaltenbrunner, A., Meier, K., Kraus, D.: Integración de redaccciones en Austria, España y Alemania: modelos de convergencia de médios. Anàlisi **38**, 173–198 (2009). Retrieved from: http://www1.Folha.uol.com.br/Folha/brasil/ult96u738633. shtml

Gradim, A.: Os géneros e a convergência: o jornalista multimédia do século XXI. In: Agora. Net#2 (s/d). Retrieved from: http://labcom-ifp.ubi.pt/files/agoranet/02/gradim-anabela-generos-convergencia.pdf

Iosifidis, P.: Global Media and Communication Policy. An International Perspective. Palgrave Macmillan, London (2013)

Jenkins, H.: Cultura da convergência. Aleph, São Paulo (2006)

Kawamoto, K.: Digital Journalism. Emerging Media and the Changing Horizons of Journalism. Rowman and Littlefield, New York (2003)

Larrondo, A., Domingo, D., Erdal, I.J., Masip, P., van den Bulck, H.: Opportunities and limitations of newsroom convergence. Journalism Stud **17**(3), 277–300 (2016). https://doi.org/10.1080/ 1461670X.2014.977611

Palacios, M.: Jornalismo online, informação e memória: apontamentos para debate (s/d). Retrieved from: http://labcom-ifp.ubi.pt/files/agoranet/02/palacios-marcos-informacao-memoria.pdf. Agora.Net#2

Poder 360: Jornais no Brasil perdem tiragem e venda digital ainda é modesta (2019, November 26). Retrieved from: https://www.poder360.com.br/midia/jornais-no-brasil-perdem-tiragem-impressa-e-venda-digital-ainda-e-modesta

Reuters institute for the Study of Journalism: Reuters Institute Digital News Report (2014). Retrieved from: Reuters_Institute_Digital_News_Report_2014_-_Spain-libre.pdf

Reuters (2019). Retrieved from: http://uk.reuters.com

Rosa, B.: O Globo é o jornal que mais cresceu em 2018. O Globo (2019, January 25). Retrieved from: https://oglobo.globo.com/economia/o-globo-o-jornal-que-mais-cresceu-em-2018-23400125

Salaverría, R., Negredo, S.: Periodismo integrado. Convergencia de medios y reorganización de redacciones. Sol90, Barcelona (2008)

Soria, C.: El periodismo entre la extinción y el renacimiento. Speech Communication Department, University of Brasília, Brazil (2011)

Thaïs de Mendonça Jorge, Post-Doctorate at the University of Beira Interior (Portugal, 2020) and the University of Navarra (Spain, 2009–10). Ph.D. in Communication and Master in Political Science by the University of Brasilia (UnB). Graduated in Communication from the Federal University of Minas Gerais, she began her teaching career in Rio de Janeiro and is currently professor at Communication Faculty/ UnB. She served as head of the Journalism Department, organized the Laboratory for the Study of Languages on Mobile Devices (Labdim) and coordinated the Postgraduate Course. She was also the Secretary of Communication to the Dean of UnB, Márcia Abrahão, and worked as a communication consultant for Unesco, the Ministry of Foreign Relations, the Inter-American Development Bank, the World Bank, and the National Institute of Educational Studies and Research Anísio Teixeira (Inep). She was academic coordinator of the Multimodal Media Project in Newsrooms, a partnership with Brunel University London, under the guidance of professor George Ghinea, and had the honour to be a visiting researcher at the Computer Science Department from the same University. Research areas: Journalism, Journalism Techniques, Digital Journalism, Media And Politics, Media Criticism, Journalism Ethics, Public Relations, Gender Studies in Newsrooms. Many of her articles were published in qualifying scientific journals. She has published six technical books in journalism. As a journalist, Thaïs worked for the most important media organizations in Brazil.

Benedito Medeiros Neto, Post-Doctorate/Informatics: Semantic Framework for Journalism by CIC/IE/UnB (2018). Post-Doctorate: Digital Literacy and Mobile Learning by the School of Communication and Art/USP (2014). Ph.D. in Information Science: Evaluation of Digital Inclusion programs, by FCI/UnB (2012). Master in Operational Research / Graph Theory by EST/ UnB (1981). Specialist in Electrical Engineering / Artificial Intelligence by UnB (1986). Electrical/ Telecommunications Engineer by UnB (1975). Visiting Professor at Computer Science Department, Brunel University, London/UK, May 2018. Project Scholar /MEC/MCTI/CAPES/ CNPq/FAPs No. 09/2014. Researcher and Professor at UnB/IE/CIC and FAC/UnB. Associate Researcher at Escola do Futuro\USP (2014-). Consultant/Evaluator of FAPESB/BA. Reviewer at IGI Global. Associate of ASSOCIAÇÃO PROFISSÃO JOURNALISTA (2019). Participant of the GT01/ENANCIB; SIMEDUC/UNIT/Aracaju; Ibero-American Magazine of CI/Faculty of Information Science/UnB. PROFESSIONAL LIFE: Director of Innovation and Development at IBrTec (2019-); At Ministry of Communications: Consultant for Digital Inclusion; Coordinator of Knowledge Management and Evaluation of the GESAC Program (2012). At ECT he was Director Manager (2002), Advisor to the Vice Presidency (1999), Advisor/Technical Support (FAT) to the Directorate of Technology and Infrastructure (1998) and Senior System Analyst (2007). He was Head of Telecommunications Section of the Telebrás System (1978). He was Professor at ESAP/ ECT (1988), Professor at CEUB/Brasília. DEVELOPMENT AND RESEARCH AREAS: Computer Science, Information and Communication; Network Engineering; ICT teaching; Informatics and Society; Collaborative Systems and Web; Semantic Web; Digital Inclusion; Digital Cities; Competence in Information, Social Networks and Evaluation of Innovation Programs. CNPq RESEARCH GROUPS: (a) JorTec/SBPJOR; (b) Journalism and Memory in Communication; (c) Technology and Digital Narratives); (d) Competence in Information.

The Practice of Data Journalism and Changes in the Professional Profile of Journalists in Newsrooms in the United States, United Kingdom, and Brazil

Suzana Guedes Cardoso

Abstract This chapter analyzes the view of Charron and De Bonville (2016) with other theoretical conceptions about Data Journalism, to help comprehend the transitions in the professional profile of journalists who work in such specific field. Agreeing that journalism, as an object, does not fit a purely theoretical model, this study aims, by focusing on the professional profile aspect, to somehow answer the following questions: What is the professional specialization that the journalist needs to have to work with data journalism? Does the journalist who works with data journalism need to have a degree or knowledge in Programming and Statistics? What is the selection process involved in organizing a team of experts? Such questions are centered on the Professional Profile aspect.

Keywords Data journalism · Journalist profile · Big data · Data visualization · Digital journalism

Foreword The emphasis on data technology has transformed productive routines in newsrooms. Traditionally, productive routines start in the development of the agenda, continue to information investigation, writing and distribution of news in different media, such as newspapers and printed magazines, radios, podcasts, television, and the web.

The phenomena of changes in journalism are studied from different angles, some dedicated to perceiving the cyclical changes, others, focused on structural changes. However, the big challenging issue is to realize how much changes in professional routines and the adoption of technologies effectively result in transformations.

We consider, as a starting point, that the transformations in journalism are a process of greater impetus and involve paradigmatic changes. Charron and De Bonville (2016), French sociologists, are authors that we rely on to reflect on the changes and transformations in journalism. In studying this phenomenon, both

S. G. Cardoso (✉)
Universidade de Brasília, Campus Darcy Ribeiro, Brasília 70910-000, Brazil
e-mail: suzana@unb.br

© The Author(s), under exclusive license to Springer Nature Switzerland AG 2022   17
B. Medeiros Neto et al. (eds.), *Digital Convergence in Contemporary Newsrooms*,
Studies in Systems, Decision and Control 370,
https://doi.org/10.1007/978-3-030-74428-1_2

resorted to the concept of the ideal type. According to Kaplan (1969), the authors use the concept of ideal type as an instrument that would make it possible to sociohistorically compare Western journalism, notably the North American one. However, the strength of the theoretical matrix founded by Charron and De Bonville led studies of changes in journalism to be also supported by both authors.

These are also the authors who, in addition to formulating the concept of the journalistic paradigm, claim that "real journalism" is an object of infinite complexity, they conclude, meaning that it is an object that does not conform "to any pure theoretical model" (Charron and De Bonville 2016, p. 40). We are especially interested in highlighting that the authors distinguish a *Normal Change*—that type of change that occurs **in** the structure or **in** the system—from a *Paradigmatic Change*—that type of change **of** the structure or **of** the system.

And, in this sense, agreeing that journalism, as an object, does not fit a purely theoretical model, as proposed by the authors, that we triangulate the view of Charron and De Bonville (2016) with other theoretical views about Data Journalism, which can help us to comprehend the transitions in the professional profile of journalists who work with data journalism: What is the professional specialization that the journalist needs to have to work with data journalism? Does the journalist who works with data journalism need to have a degree or knowledge in Programming and Statistics? What was the selection process involved in organizing the team of experts? Such questions are centered on the Professional Profile aspect.

1   Big Data, Open Data and Data Journalism

In the digital age, data that crosses the internet is electronically recorded and numerically described by zeros and ones algorithms. Analyzing and understanding the data are growing challenges among journalists familiar with the safe universe of words. Texts, photos, videos, and audios are represented in the digital network by zeros and ones. In today's world, great sources and methods of data analysis and data visualization have become increasingly important in providing information and insights that could not be obtained by traditional means of finding the news.

To Lewis (2015, p. 321), "large-scale datasets and their collection, analysis, and interpretation are becoming increasingly salient for making sense of and deriving value from digital information, writ large". Datasets are the main input of the data analysis processes. They are represented by tabular data in spreadsheet format where the lines are the records of the information, and the columns are the characteristics of these contents.

The world of data is becoming more competitive every day, as reflected in terms of volume, variety, and value. This is why we now talk about Big Data. Data is a key asset for value creation, as well as an element that favors and promotes innovation, growth, and development.

The large volume of data, called big data, involves digital datasets that are much larger than those traditionally found on print media such as magazines and newspapers. The big data has considerable variation in the size of datasets, ranging from experimental studies of small samples to large informational volumes involving census or survey data.

As Mayer-Schönberger and Cukier (2013, p. 6) describe it, "big data refers to our newfound ability to crunch a vast quantity of information, analyze it instantly, and draw sometimes astonishing conclusions from it".

Lewis and Westlund (2015) point out that big data invokes a wide range of normative claims and practical implications for journalism as a professional practice and an organizational production—from knowledge work and economic rationale to practical skills and philosophical ethics. Because of its contested nature, "big data" as a term is a messy business (Crawford et al. 2014, as cited in Lewis and Westlund 2015, p. 449), and yet it remains the most concise way of referring to a larger and complicated set of factors at play in technology and society, as well as in technology and journalism.

Howard (2014, p. 7) points out that "the use of data journalism gained momentum around the world after Tim Berners-Lee called analyzing data the future of journalism in 2010, as part of a larger conversation around opening government data up to the public through publishing it online" (according to Arthur 2010).

The big data phenomenon, also attributed by Lewis (2015), integrates the fields of social sciences, computing, and information to the Journalism area. As affirmed by the author, "these phenomena are organized around the contexts of digital information technologies being used in contemporary news work—such as algorithms and analytics, applications and automation—that rely on harnessing data and managing it effectively" (Lewis 2015, p. 321).

Lewis and Westlund (2015, p. 449) explain that

> big data is made available by the growing ubiquity of mobile devices, sensors, "smar" machines, digital trace data, digital repositories and archives, and other fragments of social and natural activity represented by clicks, tweets, likes, GPS coordinates, timestamps, and so on (for a related discussion of the "internet of things".

According to Monino and Sedkaoui (2016, p. 23),

> the world of data is becoming more competitive every day, as reflected in terms of volume, variety, and value. This is why we now speak about Big Data. Data is a key asset for value creation, as well as an element that favors and promotes innovation, growth, and development.

Big data are being used as a source of data journalism, more specifically the production of investigative reporting, making publicly available information on social and economic problems, not valued and publicized by the government.

Data journalism is part of the world movement of open data, big data and web 2.0. Journalists in various countries are transforming journalistic practices by working with open data. In this sense, journalistic practices are being determined by mass data collection technologies.

Monino and Sedkaoui (2016) explain that open data is private or public digital data. It is produced by collective bodies or—possibly outsourced—public services. The authors reiterate that it is disseminated in a manner structured accordingly to a given method and with an open license that guarantees free access to it. They also recall the possibility for anyone to reuse it without technical, legal or financial restrictions.

As stated by Monino and Sedkaoui (2016), Open Data is comprised of several sources and types of input: public data or information coming from the public sector; data from scientific research, in particular, from publicly funded research, or private sector data, which can be made public with the right incentives and privacy protections. Therefore, Monino and Sedkaoui (2016) conclude that Open Data means making this data available for access, exploitation, and reuse by any interested actor (companies, scientists, etc.).

Vallance-Jones and McKie (2017) state that Open Data has become a big activity for government and is often tied in with ideas of making government more open and transparent, what is often referred to as open government.

Data journalism has social importance when accessing, for example, open data in the creation and updating of government information of public interest. The difference between data journalism and traditional journalism consists of the new possibilities of combining the lighthouse for the news with respect to a source for the large-scale variety of digital information available on digital networks.

Vallance-Jones and McKie (2017, p. 9) explain that "data journalism is still journalism, even if the tools are computer and software, instead of notebooks and pens". They "explore how to tell the stories with data and how to combine the results of data analysis with traditional reporting" (Vallance-Jones and McKie 2017, p. 9).

Vallance-Jones and McKie (2017) point out that open data is not the same as big data. Big data is data that is so larger in volume or velocity (the speed with which is collected and transmitted) that traditional desktop analytical techniques are not sufficient to make sense of it. The authors emphasize that with specialized techniques, such as many computers working together in what is called "distributed computing", important insight can be gained into people's behaviors and interests.

Vallance-Jones and McKie (2017) explain that open datasets are generally more limited in size, from a few thousand to a few million rows of data. And despite being a large volume of data, there is software on the market for analyzing this data

2   Data Mining and Data Visualization Methods

Working in collaboration with civil society groups and governments, such journalists are employing new, increasingly available, digital data-gathering tools to tell stories through powerful interactive graphics. As a result, the disclosure of facts will increase transparency by strengthening society and its communities.

The application of complex data mining and visualization methods open space for journalistic investigations, previously unimaginable. The two procedures consist of management tools and information analysis of a set of digital data of great proportions and complexity.

As reported by Howard (2014, p. 14), "powerful web-based tools for scraping, cleaning, analyzing, storing, and visualizing data have transformed what small newsrooms can do with limited resources. Howard (2014, p. 14) argues that the embrace of open source software and agile development practices, coupled with a growing open data movement, has breathed new life into traditional Computer-Assisted Reporting (CAR). Mair et al. (2014) enunciate that Computer-Assisted Reporting date as far back as 1989, with the setting up of the National Institute for Computer-Assisted Reporting and it has been long before social media, web 2.0 and even internet.

Collaboration across newsrooms and a focus on publishing data and code that show your work differentiate the best of today's data journalism from the CAR of decades ago. Howard (2014, p. 14) points out that traditionally, Computer-Assisted Reporting focused on gathering and analyzing data as a means to support investigations decades ago. Where traditional CAR focused on analysis, the data-driven journalism of today includes data publishing, reuse, and usability. Gray et al. (2012) explain that currently, some argue that there is a difference between CAR and data journalism. Today, they say that CAR is a technique for gathering and analyzing data as a way of enhancing (usually investigative) reportage, whereas data journalism pays attention to the way that data sits within the whole journalistic workflow. In this sense, data journalism pays as much—and sometimes more—attention to the information itself, rather than using it simply as a means to find or enhance stories.

Gray comments that becoming knowledgeable in searching, cleaning, and visualizing data is transformative for the profession of information gathering, too. As written,

> journalists who master this will experience that building articles on facts and insights is a relief. Less guessing, less looking for quotes; instead, a journalist can build a strong position supported by data, and this can affect the role of journalism greatly.

Notwithstanding, considering all the technological apparatus for searching, cleaning, and visualizing data, De Maeyer et al. (2015, p. 435) consider journalism as a socio-discursive practice. They argue that journalism does not solely exist in the news that it produces, but also in discourse. This rhetoric is not just another symbolic layer, as it places on top of practices and news artifacts: practices and discourses exist in a mutually shaping relation.

Beyond the concern on social discourse and accuracy of facts, according to Monino and Sedkaoui (2016), faced with this volume and diversification, data journalism is essential to develop techniques to make the best use of all of these stocks, in order to extract the maximum amount of information. The authors emphasize that there are several approaches, such as "data mining" which are not new but respond to the principles of descriptive and predictive methods. This is an integral part of data analysis, especially when it is substantial. Monino and Sedkaoui (2016) add that

> Data analysis is a class of statistical methods that makes it possible to process a very large volume of data and identify the most interesting aspects of its structure. Some methods help to extract relations between different sets of data and thus draw statistical information that makes it possible to describe the most important information contained in the data in the most succinct manner possible. Other techniques make it possible to group data in order to identify its common denominators clearly, and thereby understand them better (Monino and Sedkaoui 2016, p. 90).

Used in information extraction from large databases, data mining scans a large amount of material looking for patterns, relationships and aggregate data (generating new data subgroups and new meanings). Data journalists face challenges in acquiring knowledge in the field of data sciences, more specifically data analysis. However, as will be seen in this study, some newsrooms work with multidisciplinary teams. Yet, it is noticeable that several journalists are investing in training specialized scientists to understand the operationalization of data mining and computational programming in data extraction.

Data preprocessing, data cleaning or data preparation is also a key part of data mining. Quality decisions and quality mining results come from quality data. Data are always dirty and are not ready for data mining in the real world, as explained by Kherdikar-Kurlekar and Anusuya (2011).

Qi et al. (2019) say that with the continuous deepening of the study of data mining, the application area of data mining gradually expanded, its influence also spread to the media industry. Data visualization technology has changed the traditional narrative mode, making the news become a product that is produced. It is worth mentioning that the large volume of information produced by data mining is not comprehensible to the general public. The data visualization method facilitates the analysis and interpretation of the data-based journalistic narrative. The news-building process in traditional journalism begins with coverage of the source of the event, goes on to fact verification, writing of the text and dissemination in different media, according to the profile of the target audience. In data journalism, the steps are as follows: data, filter (data mining or data scraping programs), data visualization and storytelling (journalistic narrative).

In the application of data visualization, visual resources such as graphs, maps, diagrams, and infographics are used, accompanied by short texts or captions to support the narrative process and the interpretation of information by public interest.

Data visualization is a method that uses visual communication techniques to present and contextualize information, giving logical and objective meaning to the large volume of data. Qi et al. (2019) explain that, from the perspective of communication, the accessibility of visual language helps to eliminate the barriers between information communication and communication. In the context of big data, data news based on this principle, so that the original boring news and massive data are hooked together.

Herzog (2016) explains that it is difficult to detect interesting or meaningful patterns by looking at columns and rows of data. The author says that "data analysts will often create many visualizations that can help show the data from different perspectives, or they will show a subset of the data" (Herzog 2016, p. 145).

Cairo (2016) formulated four basic principles of visualization to find the right graphic forms for storytelling: "think about the task or tasks you want to enable, or the message that you wish to convey; try different graphic forms; arrange the components of the graphic and test the outcomes" (Cairo 2016, p. 125).

Up to this point, we have placed this study within a spectrum of changes in journalism (Charron and De Bonville 2016) and have listed the range of issues present in Data Journalism (Lewis 2015; Mayer-Schönberger and Cukier 2013; Lewis and Westlund 2015; Howard 2014; among others). This review is important, as it allows us to see the entry of Data Journalism as a socio-historical phenomenon, as called by Buton (2009). Thus, when we incorporate elements of biographical research into the questionnaire that we will present next, in Sect. 3 of the chapter, we end up using one of the strategies of socio-history, as defined by Buton, in which collective biographical research is used (Buton 2009). Next, we will demonstrate the methodological path and the results of this research.

3 Methodological Path and Results

Seventeen questionnaires were applied to Brazilian and international journalists (Estadão, São Paulo, Brasil; O Globo, Rio de Janeiro, Brasil; Folha de S. Paulo, São Paulo, Brasil; Volt Data Lab, São Paulo, Brasil; Metrópoles, Brasília, Brasil; Reuters, New York, EUA; The Guardian and BBC, London, UK; La Nación, São José, Costa Rica, among others), in the period from 2017 to 2020. We have obtained 9 answers, among those, 2 BBC editors and 2 editors from Estadão de São Paulo answered the questions complementing them.

The questionnaire consisted of 17 questions that included personal data (name, profession, company name, function, country, and city), in addition to contact details, such as email address and telephone number.

The media outlets whose editors answered the questionnaire were:

Grupo Folha is one of the main media conglomerates in the country. It controls the daily and national newspaper Folha de S. Paulo, the news website www.folha.com.br, Datafolha—one of the most respected research institutes in the country—Folhapress—a news agency—among other communication companies and the management of the group logistics. Folha de S. Paulo, also known as Folha de São Paulo (or simply Folha), is a Brazilian newspaper published in the city of São Paulo and with national circulation, https://www1.folha.uol.com.br/.

O Globo is a Brazilian daily news newspaper, founded on July 29, 1925, and based in Rio de Janeiro. Of national circulation, by monthly subscription in print or digital formats. It is an integral part of Grupo Globo which includes Rádio Globo,

TV Globo, the printed newspaper O Globo, and the website Globo.com, https://oglobo.globo.com/.

Grupo Estado comprises of the newspaper O Estado de São Paulo, the website estadao.com.br, the radio station Eldorado and the agency Estadão Conteúdo. The newspaper O Estado de São Paulo, also known as Estadão, is a Brazilian newspaper published in the city of São Paulo since 1875. Estadão publishes daily the main and the latest news from Brazil and the world on political, economic, sports, culture, technology, lifestyle and international affairs among others, https://www.estadao.com.br/.

Grupo Metrópoles encompasses the Metrópoles news website, book publisher and radio service. Edited and published in Brasília, Brazil, the Metrópoles communication vehicle is entirely focused on the digital medium. It has a local circulation, as well as in the Midwest region of the country, https://www.metropoles.com/.

Volt Data Lab is an independent Brazilian journalism agency focused on investigation and data-driven studies in the media and communication sector, data visualization, developing projects in several areas, such as politics, economics, human rights, media, and technology. The agency operates in the lean startup communication segment. Volt Data Lab has a wide articulation with a network of collaborators, which allows us to activate the needed professionals for each job, https://www.voltdata.info/.

Reuters, the news and media division of Thomson Reuters, is the world's largest international multimedia news provider reaching more than one billion people every day. Reuters provides trusted national and international news on business and finance to professionals via Thomson Reuters desktops, the world's media organizations, and directly to consumers at Reuters.com and via Reuters TV, https://www.reuters.com/.

BBC is the world's leading public service broadcaster. It produces programs and services for audiences throughout the UK. BBC also provides a wide range of shows, content, and services on television, radio, and online for audiences across the UK. BBC World Service is available in more than 40 languages, https://www.bbc.com/.

The 17 questions, sent by email, covered aspects such as whether the company considers data journalism news; if data journalism is disseminated in all newsrooms, or if it is concentrated in a single editorial; which is the main object of data journalism in the company's newsroom; what software are most used to extract and clean data; whether the company has developed its own software to extract and clean information from databases; if the company uses data mining to extract information; if data mining is done by robots and, if so, what kind of professional is responsible for this programming activity; if a journalist working with data journalism needs to have a degree or knowledge in Programming and Statistics; how does the selection process in the organization team of data journalism specialists takes place; what the main products resulting from data journalism are about; what technology the company uses to visualize data; what are the main software used in data visualization; on the source of the data—TXT, Text bank or BAK database,

Relational database; external Big Data, Data science; how does internal Big Data work; how are concept maps used; which analysis tools are most used by the company—Tableau software, Access software, Google, Python, SPSS.

From the topics raised in the questionnaire, we chose 3 for analysis in the present chapter, as they are the 3 questions that focus on the professional profile theme, namely:

- *What professional specialization does the journalist need to have in order to work with data journalism?*
- *Does the journalist who works with data journalism need to have a degree or knowledge in Programming and Statistics?*
- *What was the selection process involved in organizing the team of experts?*

Observing the aspects related to the professional profile, listed above, two main results were attained after the undertaken analysis process: (i) the entry of social actors from other professional areas in the newsrooms, working together with journalists, such as statisticians and economists; (ii) we have identified the need for a specific additional and/or complementary qualification for the professional profile of the journalist who works with Data Journalism.

Below, we list and exemplify each of these 2 aspects resulting from the questionnaire analysis process.

(i) the entry of social actors from other professional areas in the newsrooms, working together with journalists, such as statisticians and economists.

Journalist Fábio Vasconcelos, former art and data editor of O Globo newspaper, for example, points out that "Much of data journalism today is done as teamwork. Journalists together with other specialists, in this case, people from programming and statistics".

The same multiplicity of profiles, which involve complementary areas other than just professional journalists, can be seen in another excerpt from Fábio Vasconcelos statement:

I had already worked with data reportage before we even created the data journalism team. The company was interested in highlighting people to think about reportage with data and I was called. The team, in this case, me and another colleague who already worked with data too, stayed at the Graphics Department, which had programmers that eventually we could request help to.

As stated by Janes Roberts, data journalist of Reuters New York, United States:

Each team member would be an investigative-minded journalist with solid reporting experience grounded in digging. Adept of unearthing facts through document – or data-driven reporting. Extremely numerate and comfortable with math and basic statistical concepts. At least moderately skilled in at least one programming language. Adept at SQL and working with large datasets. At least moderately experienced in geospatial analysis. At least keenly interested in, if not directly experienced in, machine learning. Able to conceive data visualizations.

To Daniel Bramatti, editor of Estado de S. Paulo, São Paulo, Brazil:

Part of the team was relocated from within the newsroom itself. A second step was to search the market for professionals with compatible interests and curriculum.

Paul Bradshaw, Data journalist of BBC and lecturer of Birmingham City University, UK, affirms that:

It varies a lot from team to team. The BBC England data unit was established with 2 full-time members and myself, in a third part-time advisory role. The head was from a BBC Online background, so I knew the organization and news values, but not necessarily many data journalism skills. The 2nd member was from local newspapers and had used spreadsheets, with a willingness to develop skills as he went. I filled in the gaps in helping with more advanced techniques and coding. When the head moved into a new data team, he was replaced by someone who had used spreadsheet techniques as a broadcast journalist elsewhere in the BBC. The BBC Shared Data Unit is headed up by a person who has experience of heading up another data unit. A second member has worked within the local press and BBC Online, and has experience of working with FOI requests, investigative stories that require spreadsheet work. A third comes from the BBC Linked Data Unit and understands the programming and conceptual side of data, while having less newswriting experience. In addition, the unit has 3-4 seconds from the regional press who apply for their roles and are trained in data journalism skills while working on stories in the unit. There is a desire to expand the team with a developer and/or designer.

Rafaela Lima, (M)Dados of the Metrópoles department coordinator, informed that:

We select journalists with an aptitude for numbers and then look for people who are experts or familiar with specific languages (python or R) and software programs to extract and clean the data. Here, we open the selection for several careers: we interview, in addition to journalists, statisticians, computer engineers, and data scientists.

Sérgio Spagnuolo, editor of Volt Data Lab agency, elucidated:

We are a small team of just four people. We have only done one selection process until today, to hire an analyst. It was a process in which we explored, not only the technical capabilities but also the familiarity working with data and producing original content.

(ii)   the need for a specific additional and/or complementary qualification for the professional profile of the journalist who works with Data Journalism.

With a background in Statistics and Programming, reporter Cecília Lago, from Estadão, São Paulo, Brazil states that:

in this regard, I do not know such a complete course in Brazil. I know colleagues who have studied abroad, where data journalism is much more advanced and consolidated in newsrooms. I took several free courses to reach my current level of specialization, which I consider almost intermediate.

A certain experience in programming was also triggered in the response of several journalists, as in the testimony of journalist Fábio Vasconcelos, from O Globo and the editor of O Estado de S. Paulo, Daniel Bramatti:

social sensibility to know how to ask questions; ability to think of data crossings, data analysis, and perhaps, programming, or at least its basics for doing base searches (SQL and Python). According to Daniel Bramatti, mastering spreadsheet software is essential. An important differential is knowing programming and something about data visualization.

The demand for a plural profile can also be seen in the testimony of Janes Roberts, Reuters Data Journalist, New York:

> a degree, no. None of the people on my team have a computer science or statistics degree. We are all self-taught. A data journalist needs to be numerate and adept at acquiring technical skills. All of the members of my team are self-taught. The most important background a data journalist should have is in reporting. We are reporters first and foremost, and weak reporting skills will make for weak data journalism. An aptitude for math and technical skills are helpful, as well.

Rafaela Lima, (M)Dados coordinator, understands that

> ideally, the journalist should be specialized in one of the alternatives: data extraction, data analysis, visualization. However, the degree of proficiency depends on the stage and maturity of the team. The basic prerequisite is the ease of handling numbers and basic knowledge of data scraping.

All things considered As in the study done by De Meyer et al. (2015, p. 437), in which "in their description of who and what counts in the notion of data journalism, respondents also attributed authority and know-how", the journalists responding to our questionnaire also used the emphasis on authority and knowledge. It took place in several testimonies, as in the following report by journalist Paul Bradshaw, from the University of Birmingham, UK:

> Many data journalists started in traditional journalism (or design) roles and were brought into data journalism teams when they were set up. Sometimes their roles are more focused on the management of projects, and their data skills are relatively limited (instead, other members of the team perform that role). In other cases, they have reskilled. Increasingly, however, data journalists have developed data skills as a student, or come from more technical backgrounds (e.g. programming, statistics) and demonstrate an understanding of news values. It is so hard to recruit data journalists at the moment that it would be difficult to stipulate a particular background they must have (De Maeyer et al. 2015, p. 437).

Although data mining is a data scraping method for Data Science, more specifically in the Computer Science area, with great demand in data journalism, the study in question pointed out that at Reuters and the BBC, major media outlets, are a regular practice. BBC data editor, Jane Roberts states that "we regularly query large datasets to tease out interesting facts, aggregations, and trends. We also have used some machine learning techniques to harvest information from unstructured data. Paul Bradshaw, BBC, use data mining as a way of identifying patterns in data. BBC also use spreadsheets or R to identify aspects like distribution, ranking, etc. once data has been extracted. O Estadão, O Globo, Folha de S. Paulo, Volt Data Lab and Metrópoles use the data mining method without regularity.

Thus, we can conclude that we have identified in the testimonies, signs that there are changes in Big Data journalism. We can also come to understand that such changes involve a broader socio-historical phenomenon (Buton 2009), which was also found in the studies of De Maeyer et al. (2015).

To top off, we also concluded that Big Data Journalism is indeed a change in journalism. Nonetheless, turning to what was proposed by Charron and De Bonville (2016), there are still not enough elements in the phenomenon under study, even if it is being analyzed by multiple research teams in different countries, to determine whether we are facing a *Normal Change*—that type of change that occurs **in** the structure or **in** the system—or a *Paradigmatic Change*—that type of change **of** the structure or **of** the system).

Therefore, more serial and longitudinal studies on the phenomenon of Data Journalism are necessary, in order to locate, from regular studies and with an international comparative perspective, remarks that help identify the processes of changes in journalism arising from the incorporation of Data Journalism. That is the sense for which this study brings relevant contributions, by identifying, briefly, in the application of the questionnaires, the following factors in the professional profile: (i) there is the entry of social actors from other professional areas in the newsrooms, working together with journalists, such as statisticians and economists; (ii) there is a need for an additional and/or complementary specific qualification for the professional profile of the journalist who works with Data Journalism.

Such factors indicate that Data Journalism needs to be observed within its sociohistoricity and it is no longer just a matter of incorporating new tools and software. It rather signalizes changes in the professional profile.

References

Arthur, C.: Analysing Data Is the Future for Journalists, Says Tim Berners-Lee. Guardian, 22 Nov. 2010. http://www.theguardian.com/media/2010/nov/22/data-analysis-tim-berners-lee (2010)

Buton, F.: Portrait du politiste em socio-historien: la "socio-histoire" dans les sciences politiques. In: Buton, F., Mariot, N. (eds.) Pratiques et méthodes de la socio-histoire, pp. 21–42. PUF-CURAPP, Paris (2009)

Cairo, A.: The Truthful Art: Data, Charts, and Maps for Communication. New Riders, Berkeley (2016)

Charron, J., De Bonville, J.: Natureza e transformação do jornalismo. FAC Livros, Insular, Brasília, Florianópolis (2016)

De Maeyer, J., Libert, M., Domingo, D., Heinderyckx, F., Le Cam, F.: Waiting for data journalism: a qualitative assessment of the anecdotal take-up of data journalism in French-speaking Belgium. Digital Journalism 3(3), 432–446 (2015). https://doi.org/ezproxy. brunel.ac.uk/10.1080/21670811.2014.9764

Gray, J., Chambers, L., Bounegru, L. (eds.): The data journalism handbook: how journalists can use data to improve the news. European Journalism Center. https://datajournalismhandbook. org/ (2012)

Herzog, D.: Data Literacy: A User's Guide. SAGE, Los Angeles (2016)

Howard, A.B.: The Art and Science of Data-Driven Journalism. Tow Center for Digital Journalism, Columbia University, New York. https://academic.commons.columbia.edu/; https://doi.org/10.7916/d8q531v1 (2014)

Kaplan, A.: A conduta na pesquisa: metodologia para as ciências do comportamento. EdUSP, Herder, São Paulo (1969)

Kherdikar-Kurlekar, A., Anusuya, S.: A study on role of data mining in research methodology. Indian J. Commer. Manage. Stud. **2**(3) (2011). http://www.scholarshub.net

Lewis, S.C., Westlund, O.: Big data and journalism: epistemology, expertise, economics, and ethics. Digital Journalism **3**(3), 447–466 (2015). https://doi.org/ezproxy.brunel.ac.uk/10.1080/21670811.2014.976418

Lewis, S.C.: Journalism in an Era of Big data: cases, concepts, and critiques. Digital Journalism **3** (3), 321–330 (2015). https://doi.org/ezproxy.brunel.ac.uk/10.1080/21670811.2014.976399

Mair, J., Keeble, R.L., Bradshaw, P., Beleaga, T.: Data Journalism: Mapping the Future. Abramis, Bury St Edmunds, UK (2014)

Mayer-Schönberger, V., Cukier, K.: Big Data: A Revolution that will Transform How We Live, Work, and Think. Houghton Mifflin Harcourt, Boston, MA (2013)

Monino, J.-L., Sedkaoui, S.: Big Data, Open Data and Data Development, vol. 2. Wiley, Hoboken, NJ (2016)

Qi, E., Yang, X., Wang, Z.: Data mining and visualization of data-driven news in the era of big data. Cluster Comput. **22**, 10333–10346 (2019). https://doi-org.ezproxy.brunel.ac.uk/10.1007/s10586-017-1348-8

Vallance-Jones, F., McKie, D.: The Data Journalist: Getting the Story. Oxford University Press, Oxford (2017)

Suzana Guedes Cardoso, Ph.D. Senior Lecturer of Journalism at the Department of Communication, University of Brasília. Postdoctoral researcher in Data Journalism at the Computer Science Department, Brunel University, UK, 2017. Ph.D. at the Postgraduate Communication Program, University of Brasília, 2014. Doctorate research at the Department of Information Systems, Brunel University, UK, 2012. Master in Communication Design at Pratt Institute, New York, 1995.

Analysis of the Newsroom Convergence Process Through the Integration of Media and Content: A Case Study

Benedito Medeiros Neto and Inês Amaral

Abstract This empirical research aimed to analyze the newsroom of a newspaper as the center of operation, place of daily decision-making, and where convergence and integration actions are facilitated by using information and communication technologies (ICT). The starting point of the research is the analysis of the work-flow of the newsroom at *La Nación*, a Costa Rican daily, as a case study of convergence and integration, seeking to understand its modernization initiatives and, thus, to prospect the use of ICT in the treatment of content and news production. The second step was to identify the non-functional and functional requirements of the production information flow in the newsroom, considering that they are the basis for describing the work environments and convergence model adopted. As the primary analytical tool, the newsroom workflow was used to analyze convergence processes and make some perspectives.

Keywords Newsrooms integration · Media convergence · Information flow · Collaborative systems · Digital journalism

1 Introduction

The technological revolution anticipates more resources and digital means available to people in the workplace, homes, and organizations. Technologies have already been able to increase the efficiency and effectiveness of production processes in the most diverse human activity areas. For example, the communication facilities between the work teams' components within the research group or in the work environment favor the elevation of the individuals' cognition for technical and

B. Medeiros Neto (✉)
Universidade de Brasília, Campus Darcy Ribeiro, Brasília 70910-000, Brazil
e-mail: medeirosneto@unb.br

I. Amaral
University of Coimbra, Largo da Porta Férrea, Coimbra 3000-370, Portugal
e-mail: ines.amaral@uc.pt

administrative procedures. In addition to facilitating collaboration within teams of a newspaper, it reviews processes and facilitates innovation (Belochio 2012).

The human horizon with the advent of technological artifacts and the massification of digital media, which Pierre Lévy (2007, 2014) called intelligence technologies, begins to experience every day "the convergence of globalization and the technological revolution and sets up new ecosystems of spoken and written language" (Santaella 2010, p. 63). Besides, there is the possibility of a collaboration of people in projects network working asynchronously.

Communication organizations that previously dealt with few possibilities of media and distribution channels now have digital multimodal media (MDM), and multiplatform incorporated in all stages of the production of news articles. Regardless of whether they belong to governments or not, the fact is that they experience challenges to keep news production processes viable, threatened due to factors such as the economic crisis in the US and Europe from 2007 or the change in the market itself. For this reason, companies and organizations with their main support in the advertising market seek alternatives and seek their support in their own digital and virtual advertising. Therefore, they started to arouse interest in academic research, market studies, and products and technological solutions (García Avilés et al. 2009).

There is a consensus that resources and technological innovation mean for the information industry to launch new products, streamline their production processes in newsrooms, and even increase revenue, with the support of new information systems, not just management. Their customers are experiencing the most significant migration to mobile devices. Entrepreneurs and their managers have adopted journalistic convergence, hoping to recover the advertising market. The reduction in the number of readers and subscribers to printed newspapers is an unquestionable fact (Aliaga et al. 2008). Another fact is that these technological resources do not always improve the human and social development of professionals in the work environment, despite the possibility of coordinating activities and ease of communication and the potential for cooperation of the components of a project team or workgroup (Filippo et al. 2011).

The objective of this research was to analyze the newsroom of a newspaper as the center of news production, a place of daily decision making, and where the integration process and convergence actions are facilitated by the use of Information, Communication and Technologies (ICT). The study also aimed to analyze the convergence of the editorial staff of a major newspaper in planning, content generation, and publication, using the case study as a research method (da Fonseca et al. 2018). Therefore, one of the relevant issues within journalistic organizations' work and training environments is how best to exploit information and communication technology resources. For this study, the integration processes and convergence actions are based on the collaborative coordination of individuals, means, and resources (Schwingel 2009).

More specifically, in large newsrooms, the news production process can be understood from the answers to the following research questions (Pimentel et al. 2006): (RQ1) How to facilitate communication between people and digital information processing systems? (RQ2) How to guarantee the coordination of people and material resources? (RQ3) How to enable cooperation in the journalistic production work environment to benefit the organization's collective reflexive intelligence and productivity? To answer these questions, it is necessary to go deeper into the combination of practices and knowledge of communication, information, cognitive and evaluative sciences, and almost always as a common contribution of computer science. The results of this study show how the convergence process took place, or how it is underway, and it was possible to make functional perspectives and others not so much (Jorge et al. 2016).

2   Context

The work environment of newsrooms is increasingly filled with new methods to deal with the exponential increase in the volume of information and multimodal digital media (MDM). Furthermore, new knowledge management techniques and procedures for coordinating the work environment are necessary, as well as dealing with multiple channels and media. Moreover, along with these ICTs come digital networks, ubiquitous computing, hybridism, mobility, multimodal, which incorporate fundamentals, notably from the areas of computing, information, and communication (da Fonseca et al. 2018).

The fact is that journalists and communications professionals who inhabit the newsrooms have to overcome their shortcomings in ICT and incorporate new technological skills to take ownership of the work in ways that alter their tasks or daily routines. Thus, new styles of being and acting in the space of interactions of the newsroom are full of interfaces with computers, smartphones, and other artifacts, which have multiplied in this second decade of the twenty-first Century. These devices and artifacts changed old techniques, journalists' work procedures, and routines. It also opens spaces for other professionals to work in the same environment. Journalists are now required to prepare texts quickly and more accurately, with annotations that favor their manipulation by computers (Medeiros Neto et al. 2019).

Task management, or coordination of activities in newsrooms, is not exempt from conflicts, especially if it is linked to performance supervision. Conflicts in collaborative collective writing processes on Web 2.0 are based on how content is created, edited, remixed, and judged by the same people who receive it, replicate, and distribute it. Web 3.0 and recent digital transformations bring permanent interactions, including intelligent agents or robots (Primo 2013).

These intense transformations, mainly technological convergence, were a cleavage point and motivated this research. There will be more radical changes in the configuration of the work environment and tasks of journalists. It is not advisable to reduce the number of workers involved in the news production process, such as the simple and straightforward elimination of traditional functions, such as the copy editor, and the significant reduction of others, such as writing.

Nevertheless, these simple actions will never be an innovative solution but a Cartesian measure of rationalization. Moreover, often, the manager, in the drive for convergence, may accidentally compromise the quality of the news and the collaborative atmosphere of the work environment (Larrondo et al. 2016).

The four central dimensions of journalistic convergence: the business, the technological, the professional, and the communicative, were consolidated with the work of Aliaga et al. (2008). In this chapter, the main focus is on technological convergence, but not reducing the importance of the others. The workflow design can assist in mapping journalistic work environments and then make it possible to compare the convergences. Initially, the methodological path went through the analysis of the workflow as an object of research. It was understood as a set of rules that allow the publication and management of texts, photos, files, Remove or any other document—additionally, possibilities to explore the integration of media and content in newsrooms.

3 State of Art

In the last decade of the twentieth century, studies on the concepts of convergence and integration of newsrooms have received attention from centers of competence, researchers, and the market itself, in an objective way. However, these two issues have been raised and investigated for more than three decades. Although, convergence continues to be delayed by some newspapers, notwithstanding sponsors' wishes and the need for modernization called by the communication market (Moretzsohn and Teixeira 2012).

3.1 Integration

The word integration appeared in the media in the late 1970s, when Nicholas Negroponte publicly defended, during a presentation to executives, the integration of the cinema, communication, and computer industries in the process of market and organizational convergence. Although the conjectures did not materialize in the way the author raised, it should be noted that Negroponte was one of the first researchers to reflect on the social transformations caused by the technological revolution.

Through the framework of diffusion of innovations theory, studies by Singer (2006) examined newsrooms and how a combination of technologies, products, staff, and geography among the previously distinct provinces of print, television, and online media. The fact is that integration starts in the layouts of newsrooms and continues in the databases, big data, and then through distribution channels.

The integration of digital media in newsrooms in countries such as Austria, Spain, and Germany favored the development of different integration processes, from simple changes in layouts to the construction of real production operations centers of texts, image generation, and voice integration in videos on social networks. The studies of these processes show the opportunities and limitations of the converging and the possibilities of delving into aspects of the hybridity of texts, images, audios, videos in the journalistic material. Integration may also serve the business interests of management, compete for spaces in the market and, at the same time, favor collaborative activities of journalists working together with specialists in technology and business and readers (Salaverría 2015).

In addition to the concept of integration, studies have been identified that endeavor to elucidate the different points of view, values, and interpretations of the individual (in this case, the journalist) and his working group involved with collaborative systems, such as a newsroom (Filippo et al. 2011). More recently, we found studies of information systems based on intelligent agents, approaches to and use of metalanguages, ontology engineering, and the semantic web (Heravi et al. 2012; Heravi and McGinnis 2015).

3.2   Convergence

Convergence is a common term in many areas of information technology (IT) and communication. It began to materialize in telecommunications and information technology and later in the 1970s in Brazil, and far from the media or journalistic context. Even in the 1930s, the term was used in research on British and European society and culture. However, only in the 1980s, this expression was used in studies on digital technological development applied "to the integration of text, numbers, images, sounds and various elements in the media." (Briggs and Burke 2016, p. 266).

Jenkins (2008) focused on the issue of culture and convergence processes. Shifting to the Mediterranean countries, we found research on a wide variety of cases, including convergence journalism in Portugal by Canavilhas (2012).

Similar pros and cons studies of convergences have been carried out in Europe (the United Kingdom's BBC Scotland, Spain's CCMA and EITB, Norway's NRK and Flemish-Belgian VRT). They have evolved recently (Larrondo et al. 2016). Among the benefits mentioned are the ease in programming the professionals' work activities and the use of multimodal digital media (multiplatform), and in some cases, pointed out productive results with innovation in the newsrooms.

There are studies of convergence and comparison of newsrooms in Brazil as well, with epistemological or empirical bases, such as the integration of the *O Globo* newsroom, which raised questions about journalism in the era of uncertainty (Moretzsohn and Teixeira 2012). Belochio (2012), in her thesis on journalism in a context of convergence, evaluated the contributions of the implication of multiplatform distribution in the expansion of communication contracts due to the impacts of the new *Zero Hora* devices. The comparisons of digital sites zerohora.com and washingtonpost.com identified collaborative journalism and social networks in the mainstream (Rublescki and Barichello 2013).

4 Empirical Study

The phenomenon of convergence, even in newsrooms, presupposes technological, market, cultural, and social changes (Jenkins 2008). In visits to major newspapers, it was verified how the work environment is changing due to the intensive use of ICT. This change was the basis for the proposal of the framework model. According to García Avilés et al., convergence in journalism is a process: "The descriptors are related to four essential areas of development in a media convergence process: (1) Project scope. (2) Newsroom management. (3) Journalistic practices. (4) Work organization." (2009, pp. 293).

Four dimensions of convergence affect the news production industry:

   i. Technological convergence—a more evident aspect goes beyond network transmission to combine content, languages, teams, and supports;
  ii. Business dimension—manifested in mergers, incorporations, and diversification of activities;
 iii. Professional dimension—it threatens and constrains the newsroom environment, affecting production structures;
  iv. Content convergence—journalism with multimodal digital media (multiplatform). The framework model is more linked to the first dimension without reducing importance from the others.

4.1 Research Methods

This empirical study uses qualitative, exploratory, descriptive methods. A case study approach was used to investigate and elucidate the different views, values, and interpretations of individuals in their workgroups, using a framework model and collaborative systems that support journalistic newsrooms. Mix methods were designed to conduct empirical research in a real context, i.e., newsrooms. Not all variables related to the phenomenon will be known or controlled due to the limitations of the research project (Creswell 2008; Wazlawick 2017).

4.2 Data Collection

This research is supported by on-the-spot interviews and direct and indirect observations as it consists of a descriptive, exploratory, and qualitative case study. In these surveys, two sources of evidence will be used for data collection:

(a) Survey of information available in digital media and made available by newspapers, before visits;
(b) Conducting in-depth interviews with journalists, experts, and managers during visits to collect the opinion or perception of users about systems, frameworks, and tools. For example: What are the advantages and disadvantages of each system? What do users think of the existing functionalities, and which system did they prefer to use?

Other contacts with the interviewees were made to complement information and resolve doubts about non-functional, and mainly, functional requirements.

5 Convergence Analysis Based on Newsroom Workflow

The economic crisis continues until today, joined by a second crisis more structurally related to communication in the current days in digital journalism (Jenkins 2008; Bertocchi 2014). The growth in access to new channels and the interest in other content produced by users were undoubtedly other factors that led the executives of newspaper companies to turn on the yellow light for their shareholders and controllers.

The second crisis has its origin in the arrival of information technologies (IT) to newsrooms, the plurality of distribution channels of the communication industry, the interactivity of social networks, and the massification of mobile devices in the hands of a population with a desire for free access, and then, the desire to produce digital content (Maia and Agnez 2011). In addition to those mentioned so far from using the Web's potential to produce digital content, convergence also implies the emergence of new journalistic genres specific to the Web.

This case study is carried out in a real context, the writing of a previously chosen newspaper, in which data and information were sought, such as:

(a) the information systems used and the artifacts involved in the news writing process;
(b) the systems for coordinating the people who participated in the newsrooms (age, sex, training, professional position, previous experience);
(c) the primary practices in the convergence process;
(d) other cultural aspects relevant to the interpretation of the collected data (Jorge et al. 2016).

5.1 Workflow for Newsrooms in Latin America

The research began with the knowledge of four newsrooms' production flow throughout the research project to understand and map the content management and production processes in the newsroom of major newspapers (Jorge et al. 2016). Simultaneously, were identified software, hardware, and products available in the market information systems, the management could leverage or analyze the audience, tools, and developed IT applications or customized to meet the operating or support needs (Rublescki and Barichello 2013).

To deepen their research on workflows, the researchers had the opportunity to spend almost a week in the *O Globo* newsroom in Rio de Janeiro. With an average daily circulation of 183 thousand copies, *O Globo* is the second Brazilian printed newspaper, following the popular *Super Notícia* (220,971) and soon after *Folha de S. Paulo* (175,441), according to the National Newspaper Association (2015). In the convergence model of *O GLOBO*, journalists work for the core of news, regardless of the content's destination, be it paper, website, or group communication means. Like many communication companies, Group Globo recently launched popular versions, such as the *Extra* newspaper.

During the visit of researchers from the University of Brasília in 2015, it was observed that the model adopted in the integrated newsroom was partially convergent, with the editorials of the printed newspaper in charge of producing material for the press and digital version (Maia and Agnez 2011). In the same visit to *O GLOBO*, the managers showed that the change or transposition of news from the paper to the portals happened little by little and progressed in the next two years.

Both at *O Globo* and *La Nación,* it was still possible to identify the use of social networks, both by newspaper professionals and users, collaborative learning software with computer support, information patterns, digital media management models, and content management systems (CMS). The ubiquitous communication of mobile devices were present in the distribution of news, the convergence of digital media and channels, and the media (Schwingel 2009; de Deus et al. 2018).

5.2 Perspective of Convergence and Integration

At the same time that the virtual or online world brings geographic distances closer, the possibility of a better expression of social relations in the work environment of a newsroom is also not guaranteed. Therefore, in the interviews, we sought to explore the planning, content generation, publication, or distribution, of the news. Then, the use of technologies to support teams, uses of collaborative systems was explored, and how the desired convergence model is made feasible, and alternatives to issues and difficulties pointed out during the convergence processes (Maia and Agnez 2011).

La Nación managers and journalists claimed to have realized the gain from interactions mediated by information systems, software products, or ICT artifacts. By only focusing on misuse or difficulties in appropriating such tools, they lose the possibility of perceiving so-called symbolic complaints. An evaluation of such technological artifacts can sometimes lead to an increase in conflict processes and a reduction (Primo 2013).

Although all the interviewed managers of the newspaper agree with the implementation of new production models that bring greater convergence and the growth in technological products, hardware, and software, its adoption is not as simple and immediate as the IT managers and unit managers. It is impossible to guarantee that the journalist's work is more important than the implemented method or model. However, it is possible to predict the drop in the quality of news content to keep pace with the Internet. The leader of the convergence process could choose to prevail in his point of view and maintain the pace of integration (Bertocchi 2014).

6 Case Study: *La Nación* Newspaper

The process of convergence and integration of the media (the newspapers *La Nación*, El Financiero, La Teja, and monthly magazines) started in 2007. The first stage was completed in 2011, with the move to a specially constructed building. About 400 people work in the newsroom. *La Nación* journalists were already working towards the digital-first concept (Jorge et al. 2016).

Plans or projects for the convergence and integration of the newspaper in question concern production in the workplace and were initially conducted by large consulting companies, with the support of vendors in the industry of hardware and software products operating in the communication sector.

Academic institutions and centers of competence in communication, information, and computing have mainly focused on the search for new convergence models and the implementation of new artifacts or frameworks for small problems, for which the software industry does not have a solution to offer. There are also assessments made of convergence cases using collaborative systems (García Avilés et al. 2009; Larrondo et al. 2016).

6.1 Convergence and Integration in La Nación

La Nación group publishes the largest circulation newspaper in Costa Rica, with an average of 100,000 copies/day. It is one of the few vehicles on the continent to have a well-integrated newsroom, whose investment was US$7 million.

The choice of *La Nación* to endeavor full integration puts this newspaper as a great case study, especially as the leader in the operation of a communications market in a relevant country in Central America. Costa Rica is a prosperous country with a more diversified economy than the rest of the region and a neoliberal policy. The newspaper had about 440 professionals, 270 journalists and 170 professionals from other areas (see Fig. 1).

In the case study of the newspaper *La Nación*, it was found that the first attempt at integration and convergence was relevant in 2007 and had strong resistance due to the print journalism culture at the expense of IT use. Convergence and integration were gradually moving forward and being absorbed by professionals, according to its chief executive (2016), which has participated in the convergence and integration processes.

The consulting firm Innovation directed the physical integration at *La Nación*. However, at first, there was no synergy between journalists who would share the same space, with support from specialists and share work tools such as station workspace or meeting spaces. To the question "What is the assessment of the Innovation's contribution?" replied the executive (2016) that they did not implement the complete model, they did not meet all the established requirements (recommendations).

The transformations at *La Nación* did not consider all the objectives of the first convergence model. Production routines were expected to be carried out in a cooperative and geographically distributed work environment. According to a humanistic and social conception, it could be managed using various technological supports and converging digital media (Larrondo et al. 2016). The newsroom has an integrated organization, and convergence is a developing goal (see Fig. 1).

Fig. 1 *La Nación*'s newsroom. *Source* Author's collection (2016)

The top guideline of *La Nación* today is that journalists should do their work with quality, no matter what platform it will publish. Each day, the professional must appropriate the use of ICT, trying to write his report in the best way and with content that meets the expectations of subscribers and readers, being less attached to the format and space of the printed newspaper. In a way, they try to reduce the conflict between print and digital.

6.2 *La Nación's Editorial Workflow*

As a post-modern newspaper, La Nación should search for models that are focused on writing and that make available the technical staff's collective intelligence, use of big data, and access to quality content available in the digital media the newspaper. As a networked citizen, the journalist is a prominent actor and could be better exploited. According to the theory of collective intelligence and Pierre Lévy's (2014) semantic sphere, the production process aims to facilitate the construction of creative and intercultural dialogues.

During the visit to *La Nación* newspaper, the project 'Implementación del Modelo Integrado en GN Medios' was presented by production engineer Laura Arroyo, including the integration process, structure, and new areas, operating fluids for the generation of containers.

During the visit of researchers da Universidade de Brasília (Fig. 2) to the newsroom workflow, it was possible to understand the newspaper's newsroom's management and content production processes. The process is called 'Editorial Integrated Newsroom Management' and comprises planning, the generation of content, and the publication. The workflow is divided into the main block (integrated newsroom) and the second block, which deals with *Gestación Documental*, including the Documentation Center. It follows the model of access to quality content available on the newspaper's digital media. Integrated newsroom comprises an integrated table, content tables, graphic and audiovisual journalism, IT development, and digital publishing (da Fonseca et al. 2018).

The journalists' teams count on the means of a large newsroom, which already brings gains for the teams and productivity. The intensive use of IT, such as support and collaborative systems, was not evidenced by the central manager. The advantage of having all the workforce concentrated in a single newsroom becomes a little more challenging to use jobs anywhere and anytime, something that technology today allows. Only one case of work outside the newsroom was tried.

By adopting the integration processes and simultaneously following convergence models already established in other countries, *La Nación* surpassed the first phase of digitization. And then Phase 2, as it implemented online structures. Likewise, the newspaper went through Phase 3, the physical integration of the newsroom and its facilities. Phase 4, i.e., the development of new languages for managing the production process, faced a complicated process of successes and failures, even though it was a vanguard case for Central America. Concerning Phase

Fig. 2 Researcers knows the workflow of the La Nación Newsroom. *Source* Authors' collection (2016)

5, which corresponds to the total merger of structures, integration of means, and production processes, it can be said that it is still planning and pending adaptation of the previous four phases.

For the current editor-in-chief of *La Nación*, integration in production and management activities still lacked. Besides, he added that the integration of human resources would receive more attention from the moment of the visit, as is already the case with material resources. What most worries editors are teams' lack of understanding of the new thinking of integration and convergence of the newsroom.

The newspaper's strategic plan conducted by the general management and another task group formed by seven journalists, with the help of IT specialists under review, receiving direct support from Computer Engineering specialists—is the main instrument of guidance. *La Nación*'s production engineer was responsible for preparing and monitoring the integrated and convergent writing workflow (Production Engineer 2016).

However, other unidentified concerns, such as the benefits of using software and information systems, address the issue of collaborative work environments. The issue has not yet been made explicit by editors and reporters, only by innovation and IT managers. What is certain is that the appropriation of the collaborative systems models in journalism at *La Nación* is considered a cutting edge line. However, it is not always accepted by all involved (IT Manager 2016).

The *La Nación* newspaper now seeks to increase the integration of multiple products, production, and publication platforms and considering the gains and mishaps experienced. This review seeks to reverse the workflow: Print and Online, for Print + Digital, and then for Digital Print. Still, these variables have not been detailed in the discussion process, as they have not yet reached the desired model. Nevertheless, the manager guarantees that this will happen later.

The implementation of an operational strategy was verified with the digitization of processes, production information flow, and workflow well defined in production engineering practices. The prospecting and use of new ICTs for the news production environment are anchored in fostering and supporting startup companies to develop new applications and collaborative solutions for everyday work practice. It is believed that other available technologies will soon be incorporated into the environment, such as semantic CMS and systems for collaborative work environments orchestrated by a framework (de Deus et al. 2018).

The future was not obvious for *La Nación* managers and editors, but it was already more evident for IT technicians and managers. The fact is that all newspapers in Latin America are looking for viable business models and the adoption of an efficient workflow for the newsroom. This is because the revenue from digital products and services does not support the newspaper company's operating costs today on a 1:10 basis. Although the production of content based on data and video journalism remains expensive, even with the facilities provided by new technologies, and each day more accessible. The content market is declining, and advertising remains on a slope.

7   Conclusions, and Further Research

The process of integrating the functions of journalistic newsrooms and the technological convergence of the media has received considerable attention from the IT sector, not only as offers to modernize management and in the search for new media markets. On the side of the communication companies, there are expectations of the possibility of survival of the information industry with full use of IT. It can be questionable every day that social networks and the Internet become the most important means of communication, including news. One of the sustainability factors is revenue from advertising, which, as previously stated, has migrated to the digital medium, often to repair lost revenue from printed newspapers.

As for the market or competition, there is a growing presence of new players exploring the advertising market, OTT (Over-The-Top, such as Facebook, Google, Netflix). These so-called Internet companies, which delivered text, audio, video, and other media more quickly and targeted specific audiences, use the latest technologies and therefore have an advantage over traditional newspaper companies. The newspaper in analysis forces the search of these actors for new paths, as everything seems to be a threat not entirely visible for the future.

Semantic web, HTML, CSS, programming languages, algorithms, statistics, big data, audio, video, image formats... What is the limit? In today's world, a journalist deals with a greater or lesser extent and in-depth with all these elements because they directly interfere in journalistic practice in all its stages. Both schools or colleges of communication embrace this new challenge, which may lead to a profound overhaul of their curricula or will place professionals with limited capacity or competence, dependent on a high degree of competence in other areas. It is not a matter of transforming the journalist or communicator into an expert in computer programming. However, the evidence shows that limiting these professionals to the exclusive function of a scribe can leave them confined to jobs in small organizations that have departments prepared for the primary supply needs of those who want to publish online. It is possible to observe the explosion of channels and ways of doing journalism independently and autonomously (Medeiros Neto et al. 2019).

It is expected that other researchers and research centers, in the first place, will arouse interest in these complex issues raised in this study and that the research funding agencies will encourage or support new projects on this topic. With these new researches, other approaches can be added. Secondly, the relationship between people after the integration process or about how it can be done, so that the attributions and hierarchies are more horizontal among the professionals, in a way that it does not show only a change of layout or use of ICT as a way to save on labor. A third point that deserves attention is how the organizational culture can deal with work overload without due salary compensation and problems related to the technology itself, to the acceptance by the more experienced journalists, given the adequacy of the objectives intended by managers, owners, or shareholders of newspapers (Larrondo et al. 2016).

References

Aliaga, R.S., Piqué, A.M., Negredo, S.: Periodismo integrado: convergencia de medios y reorganización de redacciones. Sol 90 (2008)

Belochio, V.D.C.: Jornalismo em contexto de convergência: implicações da distribuição multiplataforma na ampliação dos contratos de comunicação dos dispositivos de *Zero Hora*. Doctoral dissertation, Federal University of Rio Grande do Sul, Brazil (2012)

Bertocchi, D.: Dos dados aos formatos. Um modelo teórico para o design do sistema narrativo no jornalismo digital. Doctoral dissertation, University of São Paulo, Brazil (2014)

Briggs, A., Burke, P.: Uma história social da mídia: de Gutenberg à Internet. Editora Schwarcz, Companhia das Letras (2016)

Canavilhas, J.: From remediation to convergence: looking at the Portuguese media. Braz. Journalism Res. **8**(1), 7–21 (2012)

Creswell, J.W.: Qualitative, quantitative, and mixed methods approaches (2008)

da Fonseca, M., Ishikawa, E., Medeiros Neto, B., Victorino, M., Oliveira, E.C.: Ferramenta para anotação semântica de processos de negócio de uma redação jornalística. In: ONTOBRAS, pp. 239–244 (2018)

de Deus, V.S., Ishikawa, E., Oliveira, E.C., Victorino, M., Medeiros Neto, B., Groenli, T.M., Ghinea, G.: Towards a semantic-based content management system for journalistic writing. In: Proceedings of the 10th International Conference on Management of Digital Ecosystems, pp. 141–148 (2018, September)

Filippo, D., Pimentel, M., Wainer, J.: Metodologia de pesquisa científica em sistemas colaborativos. Sistemas Colaborativos **1**, 379–404 (2011)

García Avilés, J.A., Meier, K., Kaltenbrunner, A., Carvajal, M., Kraus, D.: Newsroom integration in Austria, Spain and Germany: models of media convergence. Journalism Pract. **3**(3), 285–303 (2009)

Heravi, B.R., McGinnis, J.: Introducing social semantic journalism. J. Media Innov. **2**(1), 131–140 (2015)

Heravi, B.R., Boran, M., Breslin, J.: Towards social semantic journalism. In: Sixth International AAAI Conference on Weblogs and Social Media (2012, May)

Jenkins, H.: Cultura da convergência. Editora Aleph, São Paulo (2008)

Jorge, T.M., Cardoso, S.G., Oliveira, E.C., Medeiros Neto, B.: Analysis of the integration process in newsrooms. The cases of Correio Braziliense, O Globo and La Nación. In: Proceedings of XIII Congreso de la Asociación Latinoamericana de Investigadores de Comunicación. Ciudad de México: Mexico (2016)

Larrondo, A., Domingo, D., Erdal, I.J., Masip, P., van den Bulck, H.: Opportunities and limitations of newsroom convergence: a comparative study on European public service broadcasting organisations. Journalism Stud. **17**(3), 277–300 (2016)

Lévy, P.: A inteligência coletiva: por uma antropologia do ciberespaço. Loyola, São Paulo (2007)

Lévy, P.: A esfera semântica. Tomo I – Computação, cognição, economia da informação. São Paulo: Annablume (2014)

Maia, K.B.F., Agnez, L.F.: A convergência digital na produção da notícia: Dois modelos de integração entre meio impresso e digital. http://www.mejor.com.br/index.php/mejor2011/MEJOR/paper/view/73 (2011)

Medeiros Neto, B., Ishikawa, E., Ghinea, G., Grønli, T.M.: Newsroom 3.0: managing technological and media convergence in contemporary newsrooms. In: Proceedings of the 52nd Hawaii International Conference on System Sciences (2019, January)

Moretzsohn, S., Teixeira, F.: A integração da redação de O Globo: questões sobre o jornalismo na era da incerteza. Encontro Nacional de Pesquisadores em Jornalismo, 10 (2012)

Pimentel, M., Gerosa, M.A., Filippo, D., Raposo, A., Fuks, H., Lucena, C.J.P.: Modelo 3C de colaboração para o desenvolvimento de sistemas colaborativos. Anais do III Simpósio Brasileiro de Sistemas Colaborativos, 58–67 (2006)

Primo, A.: Interações mediadas e remediadas: controvérsias entre as utopias da cibercultura e a grande indústria midiática. Interações em rede. Sulina, Porto Alegre (2013)

Rublescki, A., Barichello, E.: Jornalismo colaborativo e redes sociais no *mainstream*: estudo comparado do jornal zerohora.com e do washingtonpost.com. Rumores **7**(14), 99–118 (2013)

Salaverría, R.: Los labs como fórmula de innovación en los medios. El profesional de la información **24**(4) (2015)

Santaella, L.: A ecologia pluralista da comunicação: conectividade, mobilidade, ubiquidade. Paulus, São Paulo (2010)

Schwingel, C.: A produção de conteúdos no ciberespaço: sistemas de gerenciamento de conteúdos. Associação Brasileira de Pesquisadores em Jornalismo (SBPJor). VII Encontro Nacional de Pesquisadores em Jornalismo, Universidade de São Paulo *(USP)*, 35 (2009)

Singer, J.B.: Stepping back from the gate: online newspaper editors and the co-production of content in campaign 2004. Journalism Mass Commun. Q. **83**(2), 265–280 (2006)

Wazlawick, R.: Metodologia de pesquisa para ciência da computação, vol. 2. Elsevier Brasil (2017)

Benedito Medeiros Neto Post-Doctorate/Informatics: Semantic Framework for Journalism by CIC/IE/UnB (2018). Post-Doctorate: Digital Literacy and Mobile Learning by the School of Communication and Art/USP (2014). PhD in Information Science: Evaluation of Digital Inclusion programs, by FCI/UnB (2012). Master in Operational Research/Graph Theory by EST/UnB (1981). Specialist in Electrical Engineering/Artificial Intelligence by UnB (1986). Electrical/

Telecommunications Engineer by UnB (1975). Visiting Professor at Computer Science Department, Brunel University, London/UK, May 2018. Project Scholar /MEC/MCTI/CAPES/ CNPq/FAPs No. 09/2014. Researcher and Professor at UnB/IE/CIC and FAC/UnB. Associate Researcher at Escola do Futuro\USP (2014–). Consultant/Evaluator of FAPESB/BA. Reviewer at IGI Global. Associate of ASSOCIAÇÃO PROFISSÃO JOURNALISTA (2019). Participant of the GT01/ENANCIB; SIMEDUC/UNIT/Aracaju; Ibero-American Magazine of CI/Faculty of Information Science/UnB. PROFESSIONAL LIFE: Director of Innovation and Development at IBrTec (2019–); At Ministry of Communications: Consultant for Digital Inclusion; Coordinator of Knowledge Management and Evaluation of the GESAC Program (2012). At ECT he was Director Manager (2002), Advisor to the Vice Presidency (1999), Advisor/Technical Support (FAT) to the Directorate of Technology and Infrastructure (1998) and Senior System Analyst (2007). He was Head of Telecommunications Section of the Telebrás System (1978). He was Professor at ESAP/ ECT (1988), Professor at CEUB/Brasília. DEVELOPMENT AND RESEARCH AREAS: Computer Science, Information and Communication; Network Engineering; ICT teaching; Informatics and Society; Collaborative Systems and Web; Semantic Web; Digital Inclusion; Digital Cities; Competence in Information, Social Networks and Evaluation of Innovation Programs. CNPq RESEARCH GROUPS: (a) JorTec/SBPJOR; (b) Journalism and Memory in Communication; (c) Technology and Digital Narratives); (d) Competence in Information.

Inês de Oliveira Castilho e Albuquerque Amaral Associate Professor and Director of the Undergraduate Program Studies in Journalism and Communication at the Faculty of Arts and Humanities of the University of Coimbra. PhD in Communication Sciences at the University of Minho. Researcher at the Centre for Studies in Communication and Society and Associate Researcher at the Centre for Social Studies of the University of Coimbra. Coordinator of the Journalism and Society section of the Portuguese Association for Communication. Principal Investigator of the funded project "My Gender—Mediated young adults' practices: advancing gender justice in and across mobile apps". Member of the teams of the funded projects "SMaRT-EU", "(De)Coding Masculinities: Towards an enhanced understanding of media's role in shaping perceptions of masculinities in Portugal", "Opportunities and Challenges of Journalism in Open Environments". Consultant of the funded project "(De)Othering: Deconstructing Risk and Otherness". She is also a member of the Portuguese research team of the Global Media Monitoring Project 2020. Inês was an Invited Scientist Fellow at Universidad Carlos III de Madrid within the founded project ENCAGE-CM, and an Invited Professor at the University of Cape Verde. Her main research areas include audiences and media consumption; social networks, participation and social media; media and digital literacy; gender and media. She is a certified trainer by the Ministry of Education, the Union of Portuguese Journalists and the National Scientific-Pedagogical Council of Continuous Education. Inês is a member of the Cyberjournalism Observatory and Media, Information and Literacy Observatory.

Technological Convergence, Integration of Media in the Offers, and Information Workflow at the Daily *O Globo*, in Rio de Janeiro

Benedito Medeiros Neto

Abstract The purpose of this chapter is to identify the use of ITC from a case study that mapped part of the work environment of a newsroom, and to prospect the future for technologies in newsrooms. Based in Rio de Janeiro, Brazil, *O Globo* showed results of how pros and cons are balanced in collaborative processes supported by IT. It uses the 4C Collaboration Model with its dimensions of communication, coordination, cooperation and connection, as a way to present the use of workflow tools to express news production for the near future. The case study was the research method used to assess the newspaper's newsroom. The main focus was on analyzing the newsroom workflow to understand and prospect the convergence and integration process, seeking to understand the modernization and treatment of content and the production of news through the use of multiplatforms. The newsroom is changing the operation center, a place for day-to-day decision-making, and where convergence and integration actions are implemented, into an information control and data supervision center.

Keywords Convergence of newsrooms · Collaborative systems · Workflow · *O globo* · Web journalism

1 Introduction

Newsrooms' workflows are strongly impacted by the economic, professional, and technological innovations experienced in the past two decades. This implies the presence of management resources, and communication facilities within media companies, as well as permanent introduction of ICT products, technology systems and services, and new work procedures, in order to enable better results, and

B. Medeiros Neto (✉)
University of Brasília, Brasília, Brazil
e-mail: medeirosneto@unb.br

encourage collaborative work among journalists. Communication companies have been reinventing their business models to reduce losses in advertisement. Academic studies have been examining market strategies in virtual environments, and their perspectives (García Avilés et al. 2018).

Tools are used to design newsrooms' workflow and information flow, mapping thus a journalistic work environment. They also enable the comparison of convergences, and the detection of media and content integration in newsrooms. Salaverría-Aliaga and Negredo-Bruna (2013) synthesize the four central dimensions of journalistic convergence: business, technological, professional, and communicative. In this study, convergence and integration are based on the collaborative coordination of individuals, means and resources, which are mainly technological (Schwingel 2009).

Through the case study methodology, the present study aimed to analyze the convergence of the editorial staff of a major daily in the areas of planning, content generation, and publication. The study checked, in particular, whether or not technological resources are impacting the work environment. Such resources affect especially the coordination of activities, enhancing cooperation in team components, and in collective projects, within their work routine (Filippo et al. 2011; Jorge et al. 2016).

The case study mapped part of the work environment of *O Globo*,[1] a major daily newspaper in Rio de Janeiro, Brazil, to identify the use of ICT, with particular attention to collaborative technologies. From the results of this study, we expected to create a future perspective for the newsroom. The results show how collaborative processes supported by IT take place in the dimensions of communication, coordination, cooperation, and connection, based on the 4C Collaborative Model (Medeiros Neto et al. 2019). Moreover, the research at *O Globo* presents perspectives for the near future regarding news production. The Case Study Research Methodology was applied to assess the newspapers' newsroom.

As the center of activities, the newsroom is a place of daily decision-making, in which actions of convergence and integration are implemented and facilitated by the use of ICT. That research focused on the analysis of the newsroom workflow, by aiming to comprehend and to prospect the convergence and integration process; by attempting to understand the modernization and the treatment of content; and by studying the production of news through the use of multiplatforms. We sought to answer the following question: How can communication between people and the digital treatment of information facilitate people's work, additionally ensuring the coordination of people and material resources, and considering also the possibilities of cooperation toward a reflective collective intelligence?

---

[1] *O Globo* newsroom was visited in September 2015.

2 Theoretical Framework and Correlated Works

In this second decade of the twenty-first century, the techniques and work procedures in journalism have diversified, especially concerning reporters and editors. This process has allowed, among other issues, the elaboration of more accurate texts (Laje 2012). The new context also demands changes in the coordination of new actors present in newsrooms, such as system analysts and data scientists. Computational and resource systems are still needed in the integration, producing the possible convergence. However, the results are below the managers' expectations. The twenty-first century journalism has to develop new professional behavior and requirements, due to new challenges and work styles demanded by editors and managers. The modern journalist, being part and acting in a space of so many interactions and different interfaces with computers and several networks, is a clear-cut example of that transformation (Cordeiro and Lessa 2013).

2.1 Newsrooms and IT

News-generating units from around the world undergo constant changes in the configuration of the work environment, in internal communication procedures, and in the use of technologies. Newspapers' newsrooms are undeniably impacted by information technology (IT), whether in their business model, in the production process or in the treatment of their audiences. The presence of social networks started to favor the structural conditions for the development of collaborative works and the decentralization of the protagonism in the communicative process based on a single newsroom (Rublescki and Barichello 2013).

Due to convergence, journalistic contents are increasingly hybrid, as is the case of *O Globo*: the information is prepared in a multimedia or transmedia language, which combines elements such as texts, audios, videos, animations, infographics, and photos, among others (Nunes 2016). The space of interactions in the newsroom is chiefly composed by interfaces with computers, digital networks, ubiquitous computing, collaborative systems, hybridism, smartphone mobility, and several other devices, such as augmented and virtual reality. In the context of convergence, or rather of digital transformation, journalism assesses the contributions and implications of multiplatform distribution, in the expansion of advertisement contracts, and in the multiplicity of news distribution channels. Professors and researchers at the Federal University of Rio Grande do Norte (Maia and Agnez 2011) have identified two models of integration between printed and digital media.

The new context of newsrooms in communication vehicles—newspapers, radio stations, TV, or web channels—requires an intense flow of communication between its actors. The demand is for more agility in tasks and more productivity, without compromising the quality of the news (Renault 2013). New requirements for collaborative work are added to the online environment. Tasks are expected to be carried out not only in the same newsroom of the newspaper or TV studio, but also in the distribution of news production, which is sought in more than one place. In this process, it is crucial to find the best way for the news to reach the reader through multiple platforms (Larrondo et al. 2016).

In this new work environment, the coordination of the news production process is not immune to conflicts, which often take the place of collaborative processes of collective writing, stimulated by Web 2.0. Such conflicts are based on the fact that the content is created, edited, remixed, and judged by the same people who receive, replicate and distribute it (Primo 2013).

With Web 3.0, Big Data occupies the environment of data acquisition and storage, and People Analytics defines different audiences. Thus, news production, and its transmission to audiences, become more agile, and the production and distribution of news change once again. Digital Journalism, Ontology, Artificial Intelligence, automatic text generation, and Robotics are now likely in the collaborative environment of a newsroom (Bertocchi 2014).

2.2   The Convergence and Integration of Newsrooms

Newsrooms' structures were built to serve the functions of checking and writing texts, editing content, layout and publication. These functions are affected by the convergence and integration of technological resources. Laje (2012), in his *Ideology and News Technique*, had already pointed out in 1979 the need to review those structures. Nevertheless, the convergence and integration of newsrooms are still avoided by many newspapers in the twenty-first century. However, in the last decade, pressure in the field for up-to-date standards has no longer allowed immobilization.

Digital transformations, which are present in the most diverse areas of business activity, have also reached the news production process. This has taken place more specifically inside newsrooms of countries such as Austria, Spain, and Germany, favoring the development of convergence and integration models (García Avilés et al. 2018). Studies about these processes present the opportunities and limitations of convergences, besides broadening the possibilities of a deeper delving into hybrid texts, images, audios, and videos in multiplatform and integrated journalistic productions (Santos et al. 2019).

On the other hand, integration can serve business and communication interests in the dispute for new market share. Furthermore, at the same time, it facilitates the collaborative activities of journalists and technicians. These facts also call the attention of researchers and the IT Market (Salaverría 2015).

Multiplatform journalism oriented towards integration and convergence always seeks new business models. It can be seen in the face of crises in communication vehicles that have reached not only small and medium businesses, including startups, but also large corporations. Some newspapers have been leading the way, adjusting the implementation of their workflows in the news production processes, and making them more agile. Yet, it happens in the release of new journalistic products, for example within the BBC, *The New York Times* and globo.com. These enterprises are shaping up to a new journalism, as pointed out by researchers from the Graduate Program in Journalism at the Federal University of Paraíba (UFPB, in the Portuguese acronym).

Studies on the potential and limits of convergences and integration have been carried out in Europe (the United Kingdom's BBC Scotland, Spain's CCMA and EITB, Norway's NRK and Flemish VRT), and have been evolving (Larrondo et al. 2016). Among the detected advances are the dynamization in the coordination of the activities of the professionals involved in the process, and the automation of routines.

Some investigations seek to elucidate different points of view, values, and interpretations of the individuals (in this case, the journalists) and their working group, when involved or supported by collaborative systems (Filippo et al. 2011). More recently, there have been studies on intelligent agent systems, ontology and on the semantic web and approaches to the use of metalanguages (Heravi et al. 2012; de Deus et al. 2018).

Research carried out in Brazil focuses on the different dimensions of the newsroom convergence and in the integration phenomenon, considering drops in audience, and the competition in the advertising market. We have research with epistemological bases, such as the integration of a branch office of *O Globo*, in search of productivity in the use of multiplatforms (Moretzsohn and Teixeira 2012). Also, the implication of news distribution, and the extension of communication contracts for *Zero Hora* devices were examined (Belochio 2012).

The comparative study of zerohora.com and washingtonpost.com includes collaborative journalism and social networks (Filippo et al. 2011; Rublescki and Barichello 2013). The convergence of journalistic production is studied in *O Estado de S. Paulo*, *Folha de S. Paulo*, *Valor Econômico* and *O Globo* (Renault 2013; de Jorge 2013; Heravi and McGinnis 2015).

Finally, new models of news production must be highlighted, either by identifying new procedures using IT, in developing solutions, or in products for customers. Nowadays, those customers are also producers of information, based on the innovation already present in contemporary newsrooms (Medeiros-Neto et al. 2019). The results of this research, of a qualitative and exploratory nature, has led us to the evolution of our framework, integrating the various social media into Newsroom 3.1, on the basis of our interpretation of the Collaborative 4C model. This new dimension of the framework will provide support to work in the areas of communication, coordination, cooperation and connection of newsroom activities through social media.

3   Convergence Analysis Based on Workflow

Journalistic convergence is a multidimensional process, including insights of marketing and mass communication, based on the Information Technology (IT) and Telecommunications (TC) structures. These structures are affecting progressively the news production context, and providing an integration of tools, spaces, working methods, and languages, that were previously disaggregated. Thus, journalists elaborate their content collaboratively. Then, it is distributed across multiple platforms, using the specific languages for each of them (Salaverría-Aliaga et al. 2010).

Currently, newspapers are equipped with cutting-edge technology. This comprises software, hardware, content management systems, knowledge management, production management systems, audience monitoring, and analysis systems, in addition to a large number of IT tools and applications developed or customized to meet the operational or support needs in the newsroom (Rublescki and Barichello 2013).

In the characterization of digital media, Salaverría-Aliaga and Negredo-Bruna (2013) highlight that native publications have multiple formats, and are not configured in a common pattern. But they are also digital, and thus can be circulated in an integrated network. Therefore, the integration of newsrooms is only the most tangible element of the convergence process, but also the most complex. The integration model goes beyond job restructuring and staff reduction, bringing routine and journalistic practice to the center of the issue.

3.1   Classification and Evolution of Models

Information flows in journalistic production environments can also be presented by models. Several researchers observed the evolution of these models and their convergences (Garcia Avilés et al. 2018; Maia and Agnez 2011). Here, we present three groups capable of bringing together most of these convergence models:

a) *Physical or layout*: This convergence strategy, guided by people and information systems, is purely tactical. Information flows in production environments are prioritized and evaluated, with the occasional use of technological tools and production engineering. In this case, the news production processes are predominantly synchronous.

b) *Functional and integrated*: This convergence deals with asynchronous and synchronous functions, and there is an explicit information or logic structure. This model stands out for its focus on the implementation of multimodal digital models and multiple platforms. It is a hybrid process, combining asynchronous and synchronous aspects in the distribution.

c) *Convergence and semantics*: This convergence prioritizes the collection and distribution in a more intelligent way, developing from the input-output processes automatically. It is predominantly asynchronous and temporal, with the connection of the partner networks to the movement of communication. The processes are evidenced according to a semantic and ontology approach, with the treatment and storage of information from journalistic newsrooms based on standards and use of ontology.

The key to what is called convergence is the digital file (virtual data in a database, with the documentation center as the repository). It enables digital production to convert different contents such as texts, tables, photos, videos, and audio into a digital file, or into a common denominator. These digital objects are available on a computer network, whether local or remote, and stored in clouds, which expand the scenario from Web 2.0 to Web 3.0 (Neto et al. 2019). The growth in the use of metadata and the adoption of standards such as RDS and OWL, when dealing with ontologies, will lead to a Semantic Web sooner than later. Furthermore, both will favor collective intelligence and the reduction of opacity between human agents, computers and things (de Deus et al. 2018).

3.2 Use of Technological Tools to Express Production Process

(a) *Production Engineering in the treatment of information flow in the newsroom.* It comprises changes and transformations observed in the case study, and above all its effects on the environment in question. This has led newspapers to recruit and train human resources to implement and monitor these changes and, when necessary, anticipate demands on business opportunity or explicit billing (Laje 2012).

(b) *Tools for evaluating and analyzing a production process.* In print newspaper, journalists keep in their minds that information comes first, while TV news seek exciting scenes, sounds, and attractive images to go along with the news text or speech. However, when journalism goes to the web, it must preserve the knowledge of its production environment. It means, for example, that traditional concepts, such as the lead, cannot be overlooked in online journalism. In a nutshell, "it is essential to tell the reader quickly what is the news, and why he should continue reading the text" (Ferrari 2016).

3.3 Research Methods

The research method used the qualitative, exploratory, descriptive approach, and the case study to elucidate the different points of view, in a real context of a journalistic newsroom. The limitations of the field survey, such as time, do not allow detecting all the variables related to the object of study (Creswell 2008;

Wazlawick 2017). Data collections took place through on-the-spot interviews with specialists, managers, and journalists. The interviews aimed to collect the users' perception. Three sources of evidence were used:

(a) *Online data*, from globo.com and academic references available in digital media and offered by newspapers before the visits;
(b) *Survey of the technologies*, including products and information systems used, other then their functionality assessment;
(c) *Visiting the journalistic setting*: the newsroom in operation, the time and duration of convergence and integration projects, some practices, and other cultural aspects that are relevant to the interpretation of the data. In the planning, content generation and publication of news, technologies and conventional collaborative systems are used to support teams. In addition, these technologies facilitate the identification of hurdles, and respective solutions, during the convergence processes (Maia and Agnez 2011).

4   Results and Discussion

With an average daily circulation of 183,000 copies, *O Globo* is the second Brazilian printed newspaper, following the popular *Super Notícia* (220,971), and having *Folha de S. Paulo* (175,441) in third place, according to the National Newspaper Association (2015). In the convergence model of *O Globo*, journalists work for the core of news, regardless of the destination of the content, be it paper, website, or other vehicles of the group. Recently, like many communication companies, Group Globo has launched popular versions, such as the *Extra* newspaper (https://oglobo.globo.com/videos/como-funciona-redacao-16999117).

According to *O Globo*'s editor-in-chief (2015), Ascânio Seleme, the goals of the first convergence actions were almost always to: better explore the market; seek a new business model; face competition; launch new products for market segments; and improve future prospects. The convergence took place in several stages, and over a long period, and will continue to move forward. In the meantime, new workflows were established, as well as the flow of information, while workplaces were being reconfigured.

4.1   How the Convergence and Integration Happen

For Chico Amaral (Anecdotal information, 2015), executive multimedia editor at *O Globo*, the culture of professionals is a barrier for his team to achieve the desired high performance. The digital transformation of the newsroom to achieve convergence and integration is, in most cases, impacted by the natural and cultural resistance of most of the most experienced journalists to the use of ICT. Mr. Amaral

Fig. 1 Planning meeting at *O Globo*. *Source* Benedito Medeiros Neto (in 2015)

stated that all changes are focused on adapting the newsroom's workflow to a multifaceted model, with content distribution in line with the twenty-first century reader, (see Fig. 1). "The newsroom concentrated its efforts at the peak of production, which was the closing of the print newspaper, the major landmark of a daily routine", says he. "The print remains as the star, but now we are adapted to offer qualified information for all platforms throughout the day."[2]

The introduction of more flexible working hours, individual or in teams, and the convergence between print and digital are examples of such adaptation. The expanded use of collaborative systems to facilitate integration remains a challenge in the main newsroom of *O Globo*. It is not possible to disregard the conflict between generations. Solutions were sought through new information flows and layouts, as well as the formation of mixed teams, composed of more experienced professionals and digital natives (interview with the editor-in-chief, 2015).

The 2015 business model made the print newspaper a priority, due to greater financial return. The use of ICT introduced the concentration of media, making design more accessible to all (through professional supports such as infographics), and centralized photography, art and digital design, as well as search, rationalization and savings. Although everyone understands design, for example, a journalist is a journalist, and must write articles. Designers already use support software, and a collaborative system allows proactivity among younger journalists. The information flow in news production works on the basis of a service order system, according to journalist Ana Luciane Costa, from *O Globo* (in 2015).

---

[2]This statement (in Portuguese) is available at: http://propmark.com.br/midia/o-globo-muda-para-avancar-mais-no-mundo-digital.

4.2 The Newsroom Workflow

In this newspaper, and in almost all others, convergence reached newsrooms along with new technologies (ICT). However, although everything should be conditioned by or related to the business model, as stated by *O Globo*'s editor-in-chief. He said the main factors are the newspaper's revenue and survival. This context has become more critical recently, leading to the dismissal of a contingent of workers in the past five years (Moretzsohn and Teixeira 2012).

During the visit of researchers from the University of Brasília, in 2015, it was observed that an integrated newsroom model was adapted. Additionally, convergence was partially applied, as journalists of the print newspaper were in charge of producing material for the print and digital versions (Maia and Agnez 2011). In the same visit to *O Globo*, the managers showed that the changes or transfers of news from the paper to the portals would happen cautiously, as realized in its progression the following two years.

The newsrooms in the 1990s had their complexity and specificity. The print newspaper guided all operations, and hierarchy was more evident in the newsroom. Those traits were still very visible during the 2015 visit to *O Globo*. Currently, there is already a multiplatform and virtual-oriented newsroom. As expected, challenges have changed: the publication channels have expanded, and both work fronts and team diversity have increased; teams' diversity, interdisciplinarity, and the connection with readers also increased. Editors and managers of *O Globo* start from new assumptions in the second decade of the century.

Group Globo's constant repositioning in the Brazilian news market led to an update of IT tools, and a concern about talent management in the newsroom (either journalists or other professionals). Those who work in the commercial department also have a greater connection with the audience through social networks. Each editorial area of the newsroom has learned to deal with new information systems, and collaborative software, changing the newsroom routine. In 2015, managers said that they were prone to follow innovations and point out alternatives for the new work environment.

When mapping a first workflow for *O Globo*, we sought to understand and transmit knowledge about the newspaper in a structured way. It was done to provide a broader view of the newsroom, facilitate the understanding of researchers, workflow served with support for research of this work. But at the same time, it must be said that it does not represent the real world, in all its details. This is an initial diagram that is not intended to replace something that will take months of elaboration and validation, using Production Engineering tools and techniques. This workflow was possible with the monitoring of the workflow raised during the visit in 2015.

The concept of workflow in this chapter is used for the purpose of visualizing the work and production flows of *O Globo*, and other newspapers. By understanding these processes, it was possible to structure the information, applying meaning to them, so that the logic of the newspaper's functioning is understandable. In the experimental *Campus Multiplataforma*[3] newspaper, it was possible to advance semantic understanding according to the guidelines of the Semantic Web. Thus, the content can be structured to create environments in which software and systems can develop more agile commands requested by users.

The case study[4] on *Campus Multiplataforma* (2019) continued with the structuring of information in the online environment, meeting the need to organize data on the web and provide greater support for virtual newsrooms. After all, the work environment is no longer just one physical space: it became several places with an internet-mediated connection. The easiness for new formats also requires new skills.

4.3 Perspectives for Convergence and Integration in Contemporary Newsrooms

Journalism, broadcasting, and telecommunications are inserted in a scenario of total digital transformation. The same is true for most companies and organizations in this line of business. The audience is now heterogeneous, dispersed across various platforms and channels, and can even participate. The model is now all-for-all, the "gatekeepers" are left behind, and digital media became a hallmark of a post-masses era (Lemos 2008).

Having in mind the prospect of deepening convergence and integration, solutions chosen by newspaper managers sought innovative practices, with a strong presence of ICT. This has greater weight in the newsrooms of the United States, Europe, India and China, but there is resistance from more traditional journalists, and even by new professionals, who did not receive the proper training to master tools for processing large volumes of data, something that should be a priority in Journalism schools (Larrondo et al. 2016; Salaverría 2015).

Group Globo was looking for innovative practices and the assimilation of new techniques by their professionals, aiming at increased performance in news production. As of 2016, they sought out specialists who discussed the evolution of

---

[3]Originally known as *Campus Online*, then *Campus Multimídia*, and finally *Campus Multiplataforma*.

[4]The case study, which describes the procedures used to understand the stages of planning, production, and distribution of news in the newspaper lab, was based on the projects and reports developed by the four classes that took the course "Campus Multimídia" in the Department of Communication at the University of Brasília, in 2017–2018. Once the functioning of the a multiplatform campus was understood, the experimental newspaper went through changes, adaptations, and evolutions from 2017 to early 2019. The description of procedures will be the basis for the construction of a mind map of the paper's workflow, followed by the improvement in a concept map, and then the construction and presentation of the newspaper's ontology. See Silva Thallita (2019).

media in the digital age and the digital transformation accelerated in the following years, mainly in 2018 and 2019.

In the past decade, with the convergence, the group aimed to increase market share and reduce operating costs. The managers argued that meeting the requirements of the convergence processes allowed them to work with more channels, and more media, and strengthen the integration processes. Convergence was taking place according to four factors: processes, platform, people, and products.

Group Globo started integrating its different production centers much earlier. Thus, Infoglobo consolidated itself as a company producing multimedia content for different platforms. In the new model, the newsrooms of *O Globo*, *Extra*, and *Expresso* were unified in a multimedia newsroom, although keeping the identity of each brand, preserving the characteristics of the three newspapers, and continuing to explore different themes with different approaches. The new concept developed into a Central Table of Content Production. The principle was that unification promotes a permanent debate of ideas, and facilitates innovation.[5]

What is the goal with the integration of the papers? "Initially, we sought to consolidate the synchronous print production process," explained journalist and system analyst Maíra Carvalho, (Anecdotal information, 2015). Then, came the development in four phases of integration, to carry out the production of print and digital news. The four phases are:

(a) evaluation of the potential news of the day;
(b) construction of the content identified as a possible news item;
(c) evaluation of the news produced;
(d) formatting for print newspapers, and availability at web pages.

As an example of digital integration and transformation, Globo Play was implemented as a digital video platform. On main digital devices, the user can access the broadcaster's journalism, sports, and entertainment programming.[6]

5   Conclusion, and Recommendations

Economic factors have generated within journalistic organizations the search for new communication standards, and the development of business models capable of facing competition from new media, and of attracting their declining readership, whose habits have been impacted by the digital environment. As a consequence of such digital transmutations, the integration and convergence of news production in newsrooms is accelerated. We chose the newsroom of *O Globo* because, in its

---

[5]More information (in Portuguese) at: http://oglobo.globo.com/brasil/o-globo-extra-expresso-se-integram-em-uma-redacao-multimidia-20840004.

[6]Globo Play reached 10 million downloads on December 7, 2016. More info (in Portuguese) at: http://www.mobiletime.com.br/.

convergence process, it went through the digitalization phases; implanted online structures; reduced communication professionals; and physically integrated the newsroom, and its facilities.

Within the scope of collaborative information systems in work environments, there is an interest in facilitating communication, in coordinating tasks of the production process, and in supporting activities with friendlier software, either for teams or individuals. Audiences are closely monitored, and encouraged to participate in interaction programs that enable the participation and cooperation of readers and viewers in the production of journalistic content. In the horizon, one should rely on the application of framework models or even the construction of frameworks that favor the four dimensions of collaborative work, including social networks, as well as the use of agile development methods to build a collaborative environment based on software applications (apps).

The newsroom now produces not only journalistic news, but also raw information and even data. They are moving toward intensive use of social media[7] and, in a near horizon, adoption of semantic computational models, and use of artificial and robotic intelligence for text generation and in-depth research on the web with Big Data. Such subjects will be recurrent in research, projects under development, and academic work.

Digital transformation and convergence will soon shape, and support distributed and asynchronous newsrooms, in a new configuration of collective intelligence and virtual teams work. The central question is, how can communication organizations, especially journalistic ones, take advantage of social networks, now present in almost all social activities, and of the universalization of mobile devices, to build collaborative work environments? These factors will be at the center of the debate, and need research and deepening in the short term.

References

Belochio, V.D.C.: Jornalismo em contexto de convergência: implicações da distribuição multiplataforma na ampliação dos contratos de comunicação dos dispositivos de *Zero Hora* (2012). https://www.lume.ufrgs.br/bitstream/handle/10183/61450/000861516.pdf?sequence=1

Bertocchi, D.: Dos dados aos formatos – Um modelo teórico para o design do sistema narrativo no jornalismo digital (Doctoral dissertation, University of São Paulo) (2014)

Cordeiro, R.C., Lessa, W.D.: Fluxo de trabalho do designer e visualização de notícias: o caso do jornal *O Globo*. In: Proceedings of the 6th Information Design International Conference, 5th InfoDesign, 6th CONGIC [Blucher Design Proceedings, num. 2, vol. 1]. Recife (pp. 1261–1269) (2013)

Creswell, J.W.: Qualitative, quantitative, and mixed methods approaches (2008)

de Deus, V.S., Ishikawa, E., Oliveira, E.C., Victorino, M., Neto, B.M., Groenli, T.M., Ghinea, G.: Towards a semantic-based content management system for journalistic writing. In: Proceedings

---

[7]See Appendix on the Multimodal Digital Media (MDM) Project.

of the 10th International Conference on Management of Digital Ecosystems, pp. 141–148 (2018, September)

de Jorge, T.M.: Mutação do jornalismo: como a notícia chega à internet. Universidade de Brasília, Editora UnB (2013)

Ferrari, P.: Comunicação digital na era da participação. Editora Fi, Porto Alegre (2016)

Filippo, D., Pimentel, M., Wainer, J.: Metodologia de pesquisa científica em sistemas colaborativos. Sistemas Colaborativos 1, 379–404 (2011)

García Avilés, J.A., Carvajal Prieto, M., Kaltenbrunner, A., Meier, K., Kraus, D.: Integración de redacciones en Austria, España y Alemania: modelos de convergencia de medios (2018)

Heravi, B.R., Boran, M., Breslin, J.: Towards social semantic journalism. In: Sixth International AAAI Conference on Weblogs and Social Media (2012, May)

Heravi, B.R., McGinnis, J.: Introducing social semantic journalism. J. Media Innov. 2(1), 131–140 (2015)

Jorge, T.M., Guedes, S., Costa, E., Medeiros Neto, B.: Convergence experiences in Brazil and Costa Rica. Analysis of the integration process in newsrooms. The cases of Correio Braziliense, O Globo and La Nación. XIII Congreso de la Asociación Latinoamericana de Investigadores de Comunicación – Ciudad de México (2016)

Larrondo, A., Domingo, D., Erdal, I.J., Masip, P., Van den Bulck, H.: Opportunities and limitations of newsroom convergence: a comparative study on European public service broadcasting organisations. Journalism Stud. 17(3), 277–300 (2016)

Lemos, A.: Mobile communication and new sense of places: a critique of spatialization in cyberculture. Galáxia. Revista do Programa de Pós-Graduação em Comunicação e Semiótica. ISSN 1982-2553, (16) (2008)

Maia, K.B.F., Agnez, L.F.: A convergência digital na produção da notícia: dois modelos de integração entre o impresso e o digital. Anais do I Colóquio Internacional Mudanças Estruturais no Jornalismo. Brasília (2011)

Moretzsohn, S., Teixeira, F.: A integração da redação de O Globo: questões sobre o jornalismo na era da incerteza. ENCONTRO NACIONAL DE PESQUISADORES EM JORNALISMO, 10 (2012)

Laje, N.: Ideologia e técnica da notícia. rev. e ampl. Florianópolis: Insular (2012)

Neto, B.M., Ishikawa, E., Ghinea, G., Grønli, T.M.: Newsroom 3.0: managing technological and media convergence in contemporary newsrooms. In: Proceedings of the 52nd Hawaii International Conference on System Sciences (2019, January)

Nunes, A.C.B.: Jornalismo digital de quinta geração: as publicações para tablets em diálogo com o desenvolvimento da web. Revista Alceu 17(33), 19 (2016)

Primo, A.: Interações mediadas e remediadas: controvérsias entre as utopias da cibercultura e a grande indústria midiática. Interações em rede, pp. 13–32. Sulina, Porto Alegre (2013)

Renault, D.: A convergência tecnológica e novo jornalista. Braz. Journalism Res. 9(2), 30–49 (2013)

Rublescki, A., Barichello, E.: Jornalismo colaborativo e redes sociais no mainstream: estudo comparado do jornal zerohora. com e do washingtonpost. com. RuMoRes 7(14), 99–118 (2013)

Salaverría, R.: Los labs como fórmula de innovación en los medios. El profesional de la información 24(4) (2015)

Salaverría-Aliaga, R., Negredo-Bruna, S.: Caracterización de los cibermedios nativos digitales (2013)

Salaverría-Aliaga, R., García-Avilés, J.A., Masip, P.: Concepto de convergencia periodística (2010)

Santos, É., Medeiros, B., Lenzi, A., Ghinea, G.: Redações jornalísticas em contexto de convergência: um estudo comparativo exploratório no Brasil, na Costa Rica e na Inglaterra. Comunicação & Inovação 20(43) (2019)

Schwingel, C.: A produção de conteúdos no ciberespaço: sistemas de gerenciamento de conteúdos. SBPJor–Associação Brasileira de Pesquisadores em Jornalismo. VII Encontro Nacional de Pesquisadores em Jornalismo, Universidade de São Paulo (USP), 35 (2009)

Silva Thallita, A.: Construção ontológica: uma análise dos processos produtivos do *Campus Multiplataforma* para elaboração de um *workflow*. Monograph presented to the Department of Communication at the University of Brasília, as a partial requirement for obtaining a BA in Journalism (2019)

Wazlawick, R.: Metodologia de pesquisa para ciência da computação, vol. 2. Elsevier, Brasil (2017)

Benedito Medeiros Neto Post-Doctorate/Informatics: Semantic Framework for Journalism by CIC/IE/UnB (2018). Post-Doctorate: Digital Literacy and Mobile Learning by the School of Communication and Art/USP (2014). PhD in Information Science: Evaluation of Digital Inclusion programs, by FCI/UnB (2012). Master in Operational Research/Graph Theory by EST/UnB (1981). Specialist in Electrical Engineering/Artificial Intelligence by UnB (1986). Electrical/ Telecommunications Engineer by UnB (1975). Visiting Professor at Computer Science Department, Brunel University, London/UK, May 2018. Project Scholar /MEC/MCTI/CAPES/ CNPq/FAPs No. 09/2014. Researcher and Professor at UnB/IE/CIC and FAC/UnB. Associate Researcher at Escola do Futuro\USP (2014–). Consultant/Evaluator of FAPESB/BA. Reviewer at IGI Global. Associate of ASSOCIAÇÃO PROFISSÃO JOURNALISTA (2019). Participant of the GT01/ENANCIB; SIMEDUC/UNIT/Aracaju; Ibero-American Magazine of CI/Faculty of Information Science/UnB. PROFESSIONAL LIFE: Director of Innovation and Development at IBrTec (2019–); At Ministry of Communications: Consultant for Digital Inclusion; Coordinator of Knowledge Management and Evaluation of the GESAC Program (2012). At ECT he was Director Manager (2002), Advisor to the Vice Presidency (1999), Advisor/Technical Support (FAT) to the Directorate of Technology and Infrastructure (1998) and Senior System Analyst (2007). He was Head of Telecommunications Section of the Telebrás System (1978). He was Professor at ESAP/ ECT (1988), Professor at CEUB/Brasília. DEVELOPMENT AND RESEARCH AREAS: Computer Science, Information and Communication; Network Engineering; ICT teaching; Informatics and Society; Collaborative Systems and Web; Semantic Web; Digital Inclusion; Digital Cities; Competence in Information, Social Networks and Evaluation of Innovation Programs. CNPq RESEARCH GROUPS: (a) JorTec/SBPJOR; (b) Journalism and Memory in Communication; (c) Technology and Digital Narratives); (d) Competence in Information.

The Communication, Coordination, Cooperation, and Connection Dimensions, When Using Framework and Collaborative Systems in the Newsroom—A Case Study in the BBC London

Gheorghita Ghinea, Benedito Medeiros Neto,
Maria de Fátima Ramos Brandão, and Edison Ishikawa

Abstract This chapter seeks to identify the main functional requirements of the work environment in a journalistic newsroom, including collaborative tools and information systems in the dimensions of communication, coordination, cooperation, and connection. Moreover, we also wanted to identify software and frameworks that are used by collaborative technologies. We sought to verify how these IT tools support the management and production of journalistic content, with possible support of ontologies and Semantic Web standards. Based on a visit to the BBC's newsroom in Central London, the news production workflow was used as the main analytical tool to analyze the processes of convergence. The main focus was on the workflow in the newsroom in order to understand and explore the process of convergence and integration.

Keywords Newsroom · Cyber journalism · 4C collaborative model · User interface · CSCW · Groupware systems

---

G. Ghinea (✉)
Brunel University London, Uxbridge UB8 3PH, UK
e-mail: george.ghinea@brunel.ac.uk

B. Medeiros Neto · M. de Fátima Ramos Brandão · E. Ishikawa
Universidade de Brasília, Campus Darcy Ribeiro, Brasília CEP: 7091–900, Brazil
e-mail: medeirosneto@unb.br

M. de Fátima Ramos Brandão
e-mail: fatimabrandao@unb.br

E. Ishikawa
e-mail: ishikawa@unb.br

© The Author(s), under exclusive license to Springer Nature Switzerland AG 2022   63
B. Medeiros Neto et al. (eds.), *Digital Convergence in Contemporary Newsrooms*,
Studies in Systems, Decision and Control 370,
https://doi.org/10.1007/978-3-030-74428-1_5

1 Introduction

News production is undergoing a paradoxical break in continuity, because of the reduction in readership, the need to rethink journalistic practices, impacts of Information and Communication Technology (ICT), and outdated business models. In short, the need for renewal, which emerges with the aforementioned crisis factors, causes tensions in the execution of the actions and in the fulfillment of the functions corresponding to the news production environment (Belochio 2012).

One important factor in this respect is represented by the novel ways through which consumption of news manifests itself. Consumers search for content on any device and anywhere, and then produce content and make it available on their private or public networks. They are getting the most out of the possibilities offered by Web 3.0 and by LTE communications networks, Wi-Fi, and the quasi-ubiquitous presence of devices like tablets and smartphones, both inside and outside their homes.

One of the relevant questions within work and training environments in organizations, in this case in newsrooms, is the question of how ICT impacts and transforms these environments. Beyond the ubiquity offered by ICT, the joint consumption/production of news by users now transforms them in prosumers (producers/consumers) of content. In this context, media companies should seek a collaborative work environment for news production, and ICT can facilitate this goal by enabling increased communications, coordination and cooperation between people. Accordingly, the study described in this paper aimed at deepening the understanding role of Collaborative Systems of Computer Supported Work (CSCW) in production in newsrooms. It sought to answer the following question: How does connection between people and digital information management facilitate journalistic production and ensure the communication, coordination and cooperation of newsroom stakeholders?

Modern newsrooms face three-pronged challenges. Firstly, the newsrooms of major newspapers suffer technological transformations to exploit the competitive market of mass communication (Belochio 2012). Moreover, in newsrooms, one also witnesses the presence of integration processes as well as technological, communicative, professional and business convergence (Dailey et al. 2005). The second challenge is the need for a review of the production processes alongside the digital transformation of the company. Thirdly, how to obtain favourable results for the company in the face of the growth of social networks and their ability to create news. All these three points were the focus of this case study visit to the BBC which is described in this paper.

To achieve the goals described earlier, we undertook three principal activities, which encompass:

1. Identifying common patterns related to those collaborative activities, tasks, and processes in working groups, which should be reflected in groupware user interfaces to encourage the essential groupware functions (communication, coordination, collaboration and connection);

2. Identifying tools and functionalities into a strategy oriented to facilitate the work environment, and, towards this end, the interactions that take place in a group to facilitate collaborative workspaces; and

3. Evaluating the degree of use of each tool in the organization, among the team of journalists and for the professional individually, along the four dimensions of communications, coordination, cooperation and connection.

2 Collaborative Journalism in the Elaboration of News

The digital transformations of the news production processes, especially in newsrooms, in Austria, Spain and Germany have been shown to favor the development of convergence and integration models (Avilés et al. 2009). More recent comparative work (Larrondo et al. 2016) pointing out the pros and cons of convergence have also been carried out (examining the United Kingdom's BBC Scotland, Spain's CCMA and EITB, Norway's NRK, and Flemish-Belgian VRT). The studies of these processes highlight the opportunities and limitations of convergence, and they give rise to the possibility of using multimedia documents with texts, images, audios and videos to a greater degree in journalistic production, as well as increases in productivity due to the introduction of ICT in newsrooms. However, integration can also meet the business and communication interests in the competition of new spaces in the market and, at the same time, facilitate the collaborative activities of journalists and technicians. This fact has attracted the attention of both researchers and the IT market (Salaverria 2015).

Newsrooms are also experiencing changes in the configuration and the coordination of the work environment for journalists. Moreover, although there are management initiatives to facilitate collaboration, success is not always achieved due to increased competition. For example, there is physical discomfort in the use of the technologies in the newsrooms (mainly those that offer or require mobility) and loss of coexistence in the writing environment are just some of the observed characteristics that lead to the precariousness of the work of the newspaper (Marconde Filho 2009).

Journalism in the context of convergence and of integration is also affected by multiplatform distribution in the expansion of advertising service contracts. In fact, multiplatform journalism is always looking for new business models and has attracted the attention not only of large companies, but of small and medium-sized companies too, as well as startups, all of which have been catalysts for technological innovation (Salaverría 2015). This has impacted on journalistic organisations, making them more agile and enabling a faster launch of new journalistic products (Maia and Agnez 2011).

2.1 The Use of the 3C Collaboration Model in Journalism

Thus, the study and application of models as collaborative systems in the press domain is important because it helps to understand how collaboration is established between people and, consequently, to guide future information systems projects, so that the adjust of new functionalities into the environment increased the success of information systems projects. The models provide us an insight into when and how people work in groups, support us in analyzing these same groups, so we can select and design groupware (Fuks et al. 2011).

The existing literature identifies several approaches to minimize the efforts on developing user interfaces and groupware systems, that is, several collections of design patterns have been proposed, including those that emphasize providing communication, cooperation and coordination services, which include some requirements that have influence on collaboration in mobile journalism (Dailey et al. 2005; Salaverría et al. 2010; Undurraga 2017).

We believe that studying groupware dimensions from a different perspective can improve the experience of group members in the aspects of communication, coordination and collaboration. Towards this end, we employed three approaches: one based on the 3C Collaboration Model, the other on the 4C Collaboration Model, and the last on Web 3.0. We now highlight the main aspects of each of them, as well as their relationship with social networks.

The 3C Collaboration Model is often used in the literature to classify Collaborative Systems (CSCW or Groupware), but also in the development of collaborative system or information systems (Fuks et al. 2011). Moreover, some attempts have been made to use this model in the development of groupware engineering (Pimentel et al. 2006).

Newsrooms, both convergent and non-convergent, involve collaboration agents (humans and intelligent devices) to facilitate communication, coordination and cooperation. The 3C Model, adopted as reference in this research, analyzes collaboration in three dimensions that support the working groups, forming a triad composed of communication, coordination, and collaboration:

(i) **Communication Dimension**
    Communication facilitates commitment of team members and is the most important function of groupware since it is the medium through which messages are exchanged and information is shared (Fuks et al. 2003). Communications takes place in many ways nowadays (e.g. messaging, argumentation and negotiation) and is the first dimension of 3C model.

(ii) **The Coordination Dimension**
    Coordination is the second basic function of the 3C Collaborative Model and is done through management of journalistic teams, acting on digital multi-platforms, distribution channels and other activities. Coordination entails the management and administration of people, activities and resources in any company. Coordination is also the act of ensuring that the group journalists and technology professionals of a newspaper are working cooperatively and

efficiently to achieve a goal; this includes the distribution of tasks, news gathering, review and publication (Fuks et al. 2003).

Coordination (management) will prepare the multidisciplinary teams for the collaborative and cooperative work, through the preparation of actions (pre-articulation) and agreements between the members of each team. When executing (preferably well defined) tasks and managing the interdependencies of the teams, one must take into account that running a task in a team can affect other tasks and even the entire development process or the entire project itself. One feature of interdependence is reciprocity, which means that team elements are mutually interdependent (Molleman et al. 2004).

In the vision of a Senior Product Manager at BBC News: "It's down to each division leader to set the culture they want, there's no sort of BBC-wide culture directive about how you manage that kind of risk, or information sharing, or collaboration. We're all just expected to work well together in whatever way we find, I guess."

One of the most effective and efficient ways for a team to work is through agreements. An agreement can be total or partial with reservations or restrictions. Disagreement may be complemented by an argument or an alternative proposal. Clarification is a key factor in cooperation to a collaborative environment, once clarification enables explanations of unclear situations or problems inside or outside the team.

(iii) **The Cooperation Dimension**

Cooperation is the last dimension of 3C Collaboration Model and groups members of a team together in the shared space aiming at the accomplishment of the tasks managed by a coordinator. Individuals, in this case journalists, cooperate by producing, manipulating and organizing information, constructing and refining data and information such as documents, spreadsheets, graphics or infografics, etc. Journalists act on these tasks, preparing a daily newspaper, a news site or a weekly magazine. The group members rely on commands, standardized expressions, or colloquial communication, something that happens daily in newsrooms around the world (Fuks et al. 2003).

The 3C Collaboration Model has been shown to be useful in guiding the establishment of a Collaborative System. What happens in practice is that each version of the Collaborative System or groupware is developed focusing on one of the dimensions of the 3C model, thus, a solution communications problem is now solved or focused, soon after that coordination problems or cooperation problems will be tackled. Thus, development based on the 3C Collaboration Model "helps to predict which the size of collaboration should be observed in view of the modification of a given element of a system and helps the project in the application and analysis of the results obtained from case studies in question" (Pimentel et al. 2006, p. 61).

The cooperation dimension in collaborative environments refers to the possibility that multiple people can work together on the same task to achieve a common goal, e g., the elaboration of daily news for the printed newspaper or at any time for the Internet (Costa et al. 2014), that is, it is the deliberate and coordinated participation of the group members in order to achieve a specific goal (Pimentel et al. 2006). Cooperation almost always takes place through operations or work of a team in a shared physical (or, nowadays, even, virtual) space, for example a newsroom, to perform the tasks anywhere.

3   Application of 4C Collaboration Model

The 3C Collaboration Model, whilst having many uses and applications, has limitations when it comes to dealing with the case of multiplatform environments, or in that of projects employing social networks. In the particular case of journalism, the 3C Collaboration Model hasn't fully explained the environment of a newsroom. To address this issue, in this current section we present an enhancement to the 3C Model, namely the 4C Collaborative Model.

To meet the requirements of productive work in teams, it is necessary to identify a model for newsroom environments, which needs to incorporate a technically suitable supporting the work of management and contemporary news production. Accordingly, Clarke and Niall (2009) propose a collaborative model with a broader approach, the 4C Model, based on four main functions or dimensions: communication, collaboration, cooperation and connection. Social software, as an example of a collaborative system, can be defined as one that produces socialization environments through the Internet, such as: relationship networks, blogs, micro-blogs, and other softwares. Newspapers companies now make intensive use of these tools of connection. Thus, the research also classifies social software, based on the four main functions of the 4C Model: communication, cooperation, collaboration and connection (Fuks et al. 2011).

Collaborative *communication* systems enable traffic of content, whether by text, image, audio, video, or a combination of these. Collaborative *coordination* systems ensure completion of tasks and completion of the project. Collaborative *cooperation* systems enable content sharing and encouraging people to collaborate with others in problem solving. Finally, collaborative *connection* allows the formation of people networks to take care of content. So, in newsrooms, connection is a fundamental feature, as is the relation to the audience, where consumers of news content are interconnected via social network platforms.

Researchers point out that before social software, communication, coordination and cooperation were the basic functions of any collaborative system (Bassani et al. 2016). However, the presence of social networks in almost all human activities has resulted in a new modeling of the collaborative working environment. As a result, tools and software were grouped in four categories, reflecting the elements of communication, cooperation, coordination and connection. In an alternative attempt to model modern collaborative systems, and, based on exploratory research articles and interviews with experts, Schauer and Zeiller (2011) highlighted that the features of a collaborative system could be categorised along six dimensions:

(a) asynchronous content sharing;
(b) real-time (synchronous) content sharing;
(c) content management—CMS (de Deus et al. 2018);
(d) creation and edition of documents outside the workspace;
(e) social software; and
(f) people connection.

The journalistic companies need to use these basic functions or systems to evaluate and write articles, review news, gather news, and format the news according to the distribution platform and the type of the receiving devices (Laje 2012).

Schauer and Zeiller (2011) focus their research in the area of Enterprise 2.0 (Internet use, Intranet and extranet in companies, or better, Web 2.0), which involves the use of the social software platforms in a corporate scope, that is, articulating studies in the area of virtual or electronic collaboration and the use of the 4C approach initially proposed by Clarke and Niall (2009). Moreover, Schauer and Zeiller (2011) affirm that attributes such as teamwork, collaborating and transmitting knowledge are now seen as positive differentials in an employee, in this case the journalist that need to work at a convergent and integrated newsroom. The authors also point out that to assist in this process, there are collaborative information systems that help employees "in the different phases of social interaction within the teams: communication, coordination, cooperation/collaboration and connection" (Schauer and Zeiller 2011, p. 16).

3.1 Social Networks as a Connection (C)

The movement of media convergence has taken shape from recent technological advances, mainly from the emergence of the Internet and the digitization of information, which has resulted in journalism employing digital technologies to produce news faster. News is thus produced and made available to an audience increasingly familiar with the computer, smartphone and social networks (Marcondes Filho 2009).

Over the past 12 years, audience participation in news production has progressively been brought to the fore. One example and starting point in the case of the United Kingdom was the terrorist bombing of the London Underground and buses on July 7, 2005, where over 6500 emails were sent to the BBC newsroom. Here, members of the audience uploaded more than 1000 images and videos shot on mobile phones and added eyewitness accounts to the BBC's website, in the immediate aftermath of the explosions. This provided the organization with a richer collection of material than it would have ever been able to produce by itself. Indeed, if something similar were to happen nowadays, other channels such as Whatsapp, may well be used towards the same scope.

When we asked a Senior Data Architect at BBC News Future Media, how does the BBC measure audience engagement, and how does it respond to the audience's reactions, he pointed out:

> "The other thing that we've started to pay more attention to are some specific metrics around audience engagement. So rather than just looking at page views and browsers we've started to look at time spent, how far down the page do people read, if it's a video how far through do they play, that kind of thing."

Traditional media groups such as *The New York Times*, *The Bild Zeitung*, *The Guardian*, and the BBC have not set back in the face of social-network producing news content. They have started to have an active presence on social-network platforms (*The New York Times*, for instance, began to publish articles directly on Facebook in May 2015) and reveals another initiative of big media companies to attract the virtual audience and engage with the production of content on social networks.

3.2   The Tetrahedron as a Newsroom Framework Model

After consolidating social software, notably social networks, the connection (fourth C in the 4C model) dimension has become essential to newsrooms, very much as the three other dimensions, by "allowing people to make connections for content exchange among other things" (Schauer and Zeiller 2011, page 17). In this context, collaboration occurs when people, who have at their disposal great autonomy and responsibility with the collective, work together sharing goals and commitments while motivating themselves intrinsically (Bassani et al. 2016).

To develop our News Production Framework based on the evolution of Newsroom 1.0 to Newsroom 3.0 we extend the Interactive Knowledge Stack (IKS) (Behrendt 2012). The IKS is an open source community, whose projects are focused on building an open and flexible technology platform for a semantically enhanced CMS. With the resulting framework one can design a virtual newsroom where the news flows among the newsroom stakeholders (chief journalists, journalists, press officer, news agencies, independent journalists, citizen media etc.) in a custom-made process that adapts to the newsroom environment.

Figure 1 present our proposed Newsroom Framework Model, based on the IKS. It is structured along 4 dimensions. Apart from the three dimensions described above (3C Collaboration Model), the proposed model also comprises a connection face (Four C) of a tetrahedron (Fig. 1):

(a) The Communication Face [1, 2, 3] takes place in many ways nowadays (e.g. messaging, argumentation and negotiation). The Semantic Virtual Newsroom Interface is an interface to journalists that integrates the four dimensions of the

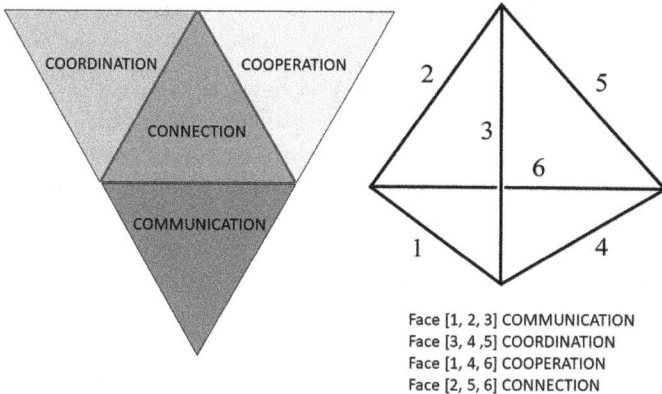

Face [1, 2, 3] COMMUNICATION
Face [3, 4 ,5] COORDINATION
Face [1, 4, 6] COOPERATION
Face [2, 5, 6] CONNECTION

Fig. 1 The four faces of the newsroom model. *Source* Authors (2017). Free translation

Newsroom Framework Model. At this user interface, journalists access the Coordination Face, Cooperation Face and Connection Face;

(b) The Coordination Face [3, 4, 5] is the WMS (Work Management System), where users model the process to produce high quality and reliable news. The workflow dimension interface, the knowledge Management Interface and the Content Management Interface permit the communication amongst them and with the Semantic Virtual Newsroom Interface (Knowledge Management Sysytem—KMS). At the end of the process, the news is available in the appropriate channel (da Fonseca et al. 2018);

(c) The Cooperation Face [2, 5, 6] is the CMS (Content Management System) and follows analogous processes contained in the KMS. So, every journalist involved in this news production has a task and an associated deadline. To help journalists in the modelling process the KMS offered semantic annotations of available processes and activities (de Deus et al. 2018);

(d) The four faces of the newsroom model (Medeiros Neto and Ishikawa 2019). The Connection Face [1, 4, 6] is a face that connects the newsroom environment to the cyberspace allowing external communications to other applications (media) such as Whatsapp, e-mail, Twitter, Facebook, Telegram and so on. These external interactions are news channels which publish customized content. They make the news available to different digital media like printed newspapers, online news, mobile news, etc. (Medeiros Neto et al. 2019).

4   Case Study: BBC London

We employed a case study methodology in order to explore news production as well as the degree of relevance of the tools/software to possible collaborative environments and tried to identify their future technological platforms. We focused on the BBC, a global major media organisation and its newsromm. Separate operational areas manage the dailing working routine at BBC studios. Among them, there are the Director-General's Office and Executive, Radio, BBC North, Finance and Business, Digital and Strategy, News Group and Television. The BBC also has three commercial subsidiaries, BBC Worldwide, BBC Studios and Post Production and BBC Global News (Franciscato et al. 2017).

Accordingly, we visited the BBC newsroom, in Portland Place, London. BBC's radio, both online and digital services have separate production teams, but they are sitting side by side in a large multimedia newsroom, aiming to share stories, videos and audios that make out a wide range of content which serves all audiences. Apart from the newsroom itself, we also visited TV studios and the web content production sections.

It was thus possible to observe aspects of inductive and operational nature (Filippo et al. 2011; Rublescki and Barichello 2013). Hereby, in this research, the analysis of the tools in the BBC newsroom was used to identify the mechanisms of collaboration to verify the potential of the tools in both physical and virtual spaces to boost the exchanges between journalists and communication professionals and, therefore, collaborative practices with the use of available technologies. Due to the limitations of the nature of the scientific enquiry undertaken, not all variables related to the real phenomena observed would have been known or controlled in the case study (Creswell 2010; Wazlawick 2009).

During the visit, we applied the methodology to explore and describe a newsroom as a way of identifying a set of technological patterns and information systems in use. The two interviewees were: Sam Ross—Senior Product Manager at BBC News and Jeremy Tarling—Senior Data Architect, BBC News Future Media. The visit to BBC on May 23th 2016 lasted 5 h and 15 min. Moreover, the research was complemented with information of qualitative and exploratory nature, provided by the institutional website and other sources such as service partners (Ishikawa and Medeiros Neto 2019).

4.1   The Collaborative Communication Dimension

The BBC newsroom in London, called "The World's Newsroom", reflects an integration (merger) under the same roof of the BBC's international journalism with state-of-the-art technologies. The space was created to facilitate multimedia production on multiple platforms: TV, radio and online. It is integrated with other BBC offices based all around the world. Communication is seen as a relevant factor for all this network.

The exchange of messages and the sharing of information are important factors to be considered in the news production process of the BBC and in any communication company. Group members must intervene in a conversation or a virtual interaction, when they need to do their tasks or when they need to seek information on virtual bulletin boards.

Collaborative communication starts by a request by a person from the newsroom, therefore, it can arise from a general or specific problem. How does one deal with group-members' spoken (verbal) requests or take care of the messages during an asynchronous or synchronous communication process? By employing the principle of usability, we contend that the answer is: Search the best interface between the information system and the real world, that is, the relation between man and machine. Table 1 shows the tools or software that can be considered in the Collaborative Communication Dimension, and their relevance in the context of the newsroom.

Since digital transformation is not only about having a faster processing or greater capacity of communication between people and things. We are already talking of the Internet of Things in which people and things will be connecting every day. The communication facilities identified in use at the BBC are already out of the newsroom differently a convectional journalistic company. Virtual meetings, by Skype for instance, at the BBC facilitate the cooperation process.

Among the communication facilities we can highlight WhatsApp in use at BBC, it is an emerging technology that works in a closed circuit for communication and also as an online social network. This particularity justifies the existence of little studies about it and that only the participants in the groups can give up the data of the conversations. The other tools or ease of communication are already well known and chosen according to the preference and culture of the organizations (Canavilhas et al. 2019).

Table 1 Artifacts, tools, and information systems for communication

Item	Name and provider	Non-functional and functional requirement	Collaborative relevance
01	Slack	Communication	Medium
02	Skype/Microsoft	Synchronous communication process, group	High
03	Whatsapp	Whatsapp (Facebook) is used quite a lot	High
04	FaceTime/Apple	Synchronous communication process, individual	Medium
05	Messenger/ Facebook	Communication	Medium
06	Snapchat	Communication plus video	Medium

4.2   Information Coordination in BBC's Newsroom

The use of ICTs aims to facilitate communication between people and digital information processing systems, ensuring the coordination of people and material resources, and the cooperation in the journalistic production workplace in order to benefit the reflexive collective intelligence. The paper focuses on the use of designed or adapted IT tools, and the application of the 3C models and then the 4C collaboration in a newsroom (Canavilhas et al. 2014; Medeiros Neto et al. 2019).

Since the BBC is a 24/7 news broadcaster, and there is always at least a channel broadcasting news, the coordination of people and material resources is continuous. Thus, there is a big studio for many channels. All BBC cameras are robotically and remotely controlled so that the camera movements are preprogrammed in an XML document, which describes a series of camera movements following a script. Therefore, according to Jeremy Tarling, there are no camera operators in the studio:

> This studio is also used in multiple activities, such as the Victoria Derbyshire program, Newsnight, the Sunday Politics or the Andrew Marr show. Thanks to that, one can observe everywhere at the studio, that all cameras are almost all the time in use. As you can see, remote-controlled lighting rigs, you know, all those visual things you expect in a TV studio.

Therefore, because all the information, of multiple activities, comes into a central area, there are always people able to receive and quickly do the triage of whatever comes in and then the information can fan out to other various channels. BBC has a layout in which the information flow (workflow) happens. Information comes into the newsroom from the outside, or rather, from elsewhere in the world. So, it might be pieces of video, audio, live streams that can be packed in the field (outside).

Another thing which is of paramount importance for the BBC according to is that "we've seen a trend over the past few years that more and more Jeremy Tarling of the audience is consuming news on devices like this (smartphone), rather than on the big screen". One can, therefore, think of this as a kind of intake information and also as information outputs, thus all the outputs are viable for the consumers and producers of information. Table 2 shows tools and software that can be used to make feasible the Collaborative Coordination Dimension and we present their relevance in the context of the BBC London.

Project management

It is up to the multidisciplinary team of professionals to prepare for collaborative and cooperative work, through the preparation of actions (pre-articulation), in the execution of tasks (insistence) and managing interdependencies, taking into account that the execution of a task affects other tasks and the whole system development process.

To this end, Sam Ross stated that: "We use Trello quite a lot. The official systems are confluence for Wiki and Jira for development management and such things as that, agile and stuff like that. In my team we like Kanban, the Kanbanize is

Table 2 Artifacts, tools, and information systems for coordination

Item	Name and Provider	Non-functional and functional requirement	Collaborative relevance
01	Trello	The ICQ styles. Technical teams will use Slack	Medium
02	Kanban (Kanbanize)	Visual management software which boosts productivity by applying lean (slim) principles to the work	Medium
03	JIRA and Wiki	Agile development management	Medium
04	ENPS/ Openmedia	The news management system used by BBC journalists to put and keep BBC news on air, is to be phased out	Very high
05	Cassandra	The Apache Cassandra database is the right choice when you need scalability and high availability without compromising performance	They don't use it, yet.
06	GraphDB	Graph data base	Low
07	Marklogic	American software business that develops and provides an enterprise NoSQL database. MarkLogic is considered a multi-model NoSQL database for its ability to store, manage, and search JSON and XML documents and semantic data (RDF triples)	Medium

easy to use visual management software which boosts your productivity by applying lean (slim) principles to your work."

A tool that was in use at the BBC was Trello, quite intuitive and easy to use, even specific tools to coordinate the production of journalistic stories. The use of Trello together Slack has been success case in many journals. Other tools already established as Wiki, and others that are emerging how the non-SQL databases.

Team organization and coordination in a journalistic project requires negotiating and assigning tasks to be carried out in a certain order and priority, in order to achieve the objectives proposed by the BBC. It should be noted the use of a specific database to store XML content, or the triple RDF. The use of graphical databases was already pointed out by BBC executives.

4.3   The Collaborative Cooperation in News Generation

The media convergence movement has taken shape from recent technological advances, especially with the emergence of the Internet and the digitization of information in newsrooms. Converged multimedia structures have been emerging since the mid-1990s, with companies around the world opting for at least some form of cooperation or synergy between previously separate employees, news-rooms, and departments. In order to examine the possibility of facilitating the journalist's work with the use of IT and enabling news production optimization, in convergent writing, it is necessary to analyze the works already developed (Fuks et al. 2011; Rublescki and Barichello 2013).

An example of Collaborative Cooperation Work from BBC readers is the publication of 455,465 pages by artists within www.bbc.co.uk/music/artists/, for example, including John Lennon, which is considered a landmark. This is much more than the staff of this public company could be able to publish, as admitted by Jeremy Tarling According to Jeremy Tarling and Sam Ross, this work collaborative is given BBC's journalists free rein to write their specific news and their newsgathering.

Collaborative Cooperation at BBC London tasks require multidisciplinary teams, either to be developed by journalists jointly (collaboratively) or individually (cooperatively) but always with a common goal. This can occur in a shared space supported by more collaborative tools and information systems shown in Table 3, which represents tools and software that can be used to support the collaborative Cooperation (Dimension).

A good example of cooperative work

Table 3 Artifacts, tools, and information systems for cooperation

Item	Name and provider	Non-functional and functional requirement	Collaborative relevance
01	Jportal	An open collaborative way of planning news across different mediums. It has mobile clients, there are lots of pockets of people that find it very useful, but most people blame the fact that it's another tool	Low (use)
02	News Lab/ Google	Is an R&D component that produces lots of different technologies to optimise workflow	Very high
03	Office/ Microsoft	We get *Office/MS* on our laptops; collaboration-wise we'll use personal gmail accounts	Medium
04	Sharepoint	CMS—the teams and partner organizations share and collaborate on content from anywhere and on any device	Medium
05	DropBox	For file storage, we use *Box.com*, but the commercial version. So, that's sort of a synchronisation tool that will have some collaboration features	High
06	Google Drive	Productivity wise, we use a lot *google docs* and things like that whenever we're trying to collaborate on work	Medium
07	Github	Repositories can contain folders and files, images, videos, spreadsheets, and data sets—anything your project needs. Tech teams use Github to store stuff	Medium
08	Ontologies	It allows us to classify our content into pieces. So pieces about a thing which is of a particular type and has a particular relation to another thing	High

Music content has the particularity of being built with information originating from three distinct sources: the BBC Music Beta. itself through reviews produced by specialist in music and culture in general; an open database called Music Brainz, which provides the complete discography of the artist; and from Wikipedia, also open, which supplies the BBC pages with the biography of the musician. Both Music Brainz and Wikipedia are platforms built on the logic of linked data, which allows dialogue with the BBC website, which has its own way of information linking.

Audience engagement

The BBC had a huge audience in May 2016, as the news website gets, for example, 14.5 million accesses by a single browser every day. "The other thing that we've started to pay more attention to are some specific metrics around audience engagement. So rather than just looking at page views and browsers we've started to look at time spent, how far down the page do people read, if it's a video, how far through do they play, that kind of thing", said Jeremy Tarling in his interview, in 2016.

Therefore, it is necessary to incentivize the use of other elements and the use of new tools. BBC for quite a long time would just measure visits of the virtual readers on pages, is what their executives said. Not only pages visits, more browsers and other things everybody already knows. The fact is that there are things that don't really tell you very much.

Feedback on cooperation

At the BBC, it is normal to contribute or solicit feedback on the presented design solutions (prototypes or documents), as well as the use of tools and systems which make this feedback process possible in the newsroom. This happens notwithstanding the fact that most of the time, the feedback was associated with discussion (through suggestions, agreement/disagreement, and questioning) throughout the development of a project.

4.4 Connection and New Media

The ubiquitous and permanent presence of ICT results in interactive connectivity nowadays. Indeed, this is the actual novelty which is generating these increased cognitive abilities of people by the use of new technologies. These technologies virtually eliminate time and space in people's interactions and until with things. All these are new frontiers that technologies can allow and going forward.

With the growing presence of people in social networks online for communication and various human activities, audience members now send small texts, images and videos shot with their own cell phones. In addition, news have eyewitnesses and in the case of the BBC website, the reader can see the sequence of events, thus providing a rich source of news material, which the BBC probably wouldn't have been able to generate through traditional means.

"We've put a recent focus on text overlays on our content, so you can watch videos without the audio. So, there's a lot of that. They have got a new format for that which was very large within a news stream sort of Facebook feed kind of format. So, it is quite chunky. I've seen square formats, I've seen gifs. I don't know where that comes from because there's certainly lots of media experiments happening", said Sam Ross in his interview.

According to Jeremy Tarling, BBC London took notice of new media early and appointed a person in charge of new media—news streams, social networks, that sort of stuff. There's an editor for the website, but there isn't an editor for a particular kind of output. There's an editor for social media, there's an editor for the website, there is an editor for other sectors of new media.

Connection and New Media at BBC London support multidisciplinary teams, either to be developed by journalists jointly (collaboratively) or individually (cooperatively) but always with a common goal. This can occur in a shared space supported by more collaborative tools and information systems shown in Table 4, which shows the tools and software that can be used for the Connection Dimension.

Reddit is a social network with news aggregation, web content rating, and a discussion website. Reddit's registered community members can submit content, such as text posts or direct links. Registered users can then vote submissions up or down to organize the posts and determine their position on the site's pages. Sam Ross said in his interview, in 2016:

Table 4 Artifacts, tools, and information systems for connection

Item	Tool name and provider	Non-functional and functional requirement	Collaborative relevance
01	Google Analyzer	Measures audience at websites. Hit, visits, etc.	Medium
02	Lean Analytics	Amazon's book. About how do you grow a small company into something that makes a lot of money very quickly	High
03	Multivariate Testing—MVT	Measure audience engagement	Medium
04	Social media, Whatsapp, Facebook, Instagram, and Twitter	How many people have come to us from social media?	Very high
05	REDDIT (https://www.reddit.com/)	It is a social news aggregation, web content rating, and discussion website. Reddit's registered community members can submit content, such as text posts or direct links	High
06	iwatch	News for smart watch, new format	Irrelevant

> Very well on Reddit. Interestingly international users share BBC.com and BBCnews.com pages more than the UK actually. American audiences when it comes to international stories, they have a distrust of their own, CNN and what have you. Not a distrust so much, but they put more validity when BBC Worldwide reports on something. So typically, when you see a news story in the global Reddit threads it will be a BBC news one—which is interesting to me.

So typically, when you see a news story in the global Reddit threads it will be a BBC news one—which is interesting to me. We hit the front page of Reddit—which is a very coveted thing to do—we're about 3 or 4 times a day.

In Brazil, things were already happening in terms of new media. Communication companies already benefit from social media all around the world in their programs. For instance, a program was broadcasted by Ricardo Senna of the BBC Brazil, which moved through the streets of Rio de Janeiro. This program was transmitted live from Monday to Friday, at 12:15 pm, during the 2016 Olympic Games on BBC Brazil page on Facebook, which is linked by the website bbcbrasil.com. In the first edition, the theme was the spirit of the Brazilians regarding the completion of the world's largest sporting event (Ishikawa and Medeiros Neto 2019).

The BBC Brazil debuted on Friday 5th August, on the very day of the 2016 Olympic Games' opening ceremony by # BBCporaí. The show featured live and daily on Facebook, with unique stories from the Olympic City. The hashtag #2BBCporaí used new software that allows real-time editing, inserts of different reporters from the BBC and public participation via social media applications such as Skype and WhatsApp. Whereas today the BBC is using 'slow news' to fight fake news. It is possible to see: https://digiday.com/uk/bbcs-slow-news-focus changing-newsroom-dynamics.

4.5 The Semantic and the Ontologies at the BBC

The use of the Labels and Metadatas in cyberjournalism. The BBC website began its tagging process in 2002, limiting itself to using keywords to optimize indexing by the portal's internal search engine and then by Google. Two years later, in 2004, metadata began to be perceived in portal as an instrument to improve the distribution and aggregation of content. The BBC then decided to use controlled vocabularies (Quintarelli et al. 2007).

The semantic content

Traditional media organizations are already embracing Semantic Web technologies, but not yet making combination with the user-generated content. The BBC is an example of a broadcasting corporation which is utilizing Semantic Web technologies with business partners, in their web sites, mainly for their programming and content of music, but not yet for news production daily (Correa and Bertocchi 2012).

The first investments of the BBC to integrate the contents in semantic form occurred in 2007. This effort was done during the creation of the BBC's programs database with the indexation of all the programs of radio and television. The use of semantic technology has favoured access to BBC content, now discovered by users in many different ways, and by service providers. The BBC News Labs implements a global innovation strategy through the design and engineering of products that facilitate journalists' work (Zaragoza-Fuster and García-Avilés 2020).

The content teams within the BBC have obtained a focal point around which they can organize the entire content of the organization, said Jeremy Tarling in his interview, 2016: MarkLogic is considered a NoSQL database for its ability to store, manage, and search documents and semantic data. It stores content, which is an XML data base, very expensive. The use of the Ontologies at BBC London.

The feasibility in the use of ontologies during the text production at newsrooms is a reality, that is, during the moment journalists are deciding which terms to use in the news, in order to enhance information disponibilization. The difference to other approaches is introduced in news production, the frequently use of the ontologies not only facilitates the publication but assures the republication and the information retrieval (Oliveira et al. 2016).

"They sought to verify how tools and systems based on information technology (IT) well support the management and production of journalistic content, and how the use of ontologies and Semantic Web standards can be introduced to the BBC", said the executives, in 2016.

This site shows open ontologies: http://www.bbc.co.uk/ontologies making available to all the access to the ontologies at the BBC, it has been used to support its audience when accessing applications such as BBC Sport, BBC Education, BBC Music, new projects and more. These ontologies were the basis of the Linked Data Platform. If someone would like to access an RDF Turtle version of an ontology, this is fairly feasible. The BBC has increased the practice of the use of the Ontological Infrastructure for a Semantic Newspaper (García et al. 2006).

5 Conclusions and Future Work

The advances identified during the visit to BBC in London are related to the perception of mutations into a more complex collaborative work environment. This process was triggered, mainly, by the introduction of Information and Communications Technology (ICT) and Semantic Web, and more recently, by digital transformations of the BBC. This consists of the use of tools and software capable of integrating the Internet with routines of the news production process and the use of Semantic Web technology, such as ontologies.

This paper briefly describes the migration of the 3C Collaboration Model to 4C (Four C) in newsrooms, and identifies the most usual technologic patterns, tools and their interfaces of the groupware, in use. The analysis and results of this research provide guidance for the managers that aim to integrate software solutions with their user interfaces in a journalistic team. Additionally, we had a question to answer, whether or not their implementation or use could contribute to improving communication, coordination, cooperation, and connection in working groups or daily routine within a newsroom.

It was observed in the visit to the facilities at the BBC in London that social networks strengthen the collaboration. Furthermore, at the same time, they behave as a new media, as can be noticed when analysing the Model 4C in the dimensions of: collaborative communication, collaborative coordination, collaborative cooperation and collaborative connection, as part of or supported by, in the work environment. This is directly influencing the way news is produced and relationships are strengthened with readers (prosumers of content). A new reality has been seen in England, which is a major focus on social media or collaborative connection to the 4C Model. The BBC registered about 85 million subscribers, in 2016. For example, Facebook had 29 million followers and Instagram had 2 million, in the same period. And then, Twitter was broken down into a Breaking News, a UK news, World News, and there were some smaller social networks. This information is part of the survey of a visit to BBC London, in May 2016.

The research identified a new journalistic procedure under development and the use of many technological tools that allow journalists and specialists to work acting in all stages of the production and distribution of content. The BBC seeks to expand the possibilities of the citizen to contribute further with content and having a critical posture, and at the time, to be connected with his community and with the world. Therefore, there are several aspects to explore as future works, such as the implementation of the proposed collection of technologic patterns into newsrooms, which will then be analyzed by means of a heuristic evaluation to verify their effectiveness from the point of view of journalism too.

Acknowledgements This research has been sponsored by the Brazilian Ministry of Education—CAPES, under the Grant Number MEC/MCTI/CAPES/CNPq/FAPs 88881.068354/2014-01.The project is part of the Experimental Laboratory for the Study of Digital Languages for Mobile Devices (Labdim) of the Faculty of Communication, University of Brasília, registered under number 485 707 in CNPQ/2013-6, in partnership with the Department of Computer Science (UnB) and Brunel University London, UK.

References

Avilés, J.A.G., Prieto, M.C., Kaltenbrunner, A., Meier, K., Kraus, D.: Integración de redacciones en Austria, España y Alemania: modelos de convergencia de medios. Anàlisi: quaderns de comunicació i cultura (38), 173–198 (2009)

Bassani, P.B.S., dos Reis, A., Dalanhol, D.: Análise da colaboração em ambientes digitais para compartilhamento de atividades de aprendizagem: uma perspectiva com base em Learning Design. In: Brazilian Symposium on Computers in Education (Simpósio Brasileiro de Informática na Educação-SBIE), vol. 27, No. 1, p. 1215 (2016)

Behrendt, W.: The interactive knowledge stack (IKS): a vision for the future of CMS. In: Semantic Technologies in Content Management Systems, pp. 75–90. Springer, Berlin, Heidelberg (2012)

Belochio, V.D.C.: Jornalismo em contexto de convergência: implicações da distribuição multiplataforma na ampliação dos contratos de comunicação dos dispositivos de Zero Hora.

Tese de Doutorado. Programa de Pós-Graduação em Comunicação e Informação da Universidade Federal do Rio Grande do Sul (2012)

Canavilhas, J., Satuf, I., Luna, D., Torres, V.: Jornalistas e tecnoatores: dois mundos, duas culturas, um objetivo. Esfera, (5) (2014)

Canavilhas, J., Colussi, J., Moura, Z.B.: Desinformación en las elecciones presidenciales 2018 en Brasil: un análisis de los grupos familiares en WhatsApp. El profesional de la información **28** (5) (2019)

Clarke, A.A., Niall, C.O.O.K.: Enterprise 2.0: how social software will change the future of work. Legal Inf. Manage. **9**(2), 148 (2009)

Correa, E.S., Bertocchi, D.: The cybercultural scene in contemporary journalism: semantic web, algorithms, applications and curation. MATRIZes **5**(2), 123–144 (2012)

Costa, A.P., Loureiro, M.J., Reis, L.P.: Modelo de Análise de Processos de Desenvolvimento de Software Educativo. Revista Lusófona de Educação **27**(27), 181–200 (2014)

Creswell, J.W.: Projeto de pesquisa: métodos qualitativo, quantitativo e misto; tradução Magda Lopes. Artmed, Porto Alegre (2010)

da Fonseca, M., Ishikawa, E., Medeiros Neto, B., Victorino, M., Oliveira, E.C.: Ferramenta para Anotação Semântica de Processos de Negócio de uma Redação Jornalística (Tool for Semantic Annotation of Business processes in a Newsroom). In: ONTOBRAS, pp. 239–244 (2018)

Dailey, L., Demo, L., Spillman, M.: The convergence continuum: a model for studying collaboration between media newsrooms. Atlantic J. Commun. **13**(3), 150–168 (2005)

de Deus, V.S., Ishikawa, E., Oliveira, E.C., Victorino, M., Neto, B.M., Groenli, T.M., Ghinea, G.: Towards a semantic-based content management system for journalistic writing. In: Proceedings of the 10th International Conference on Management of Digital EcoSystems, pp. 141–148. ACM (2018)

Filippo, D., Pimentel, M., Wainer, J.: Metodologia de pesquisa científica em sistemas colaborativos. Sistemas Colaborativos, 1, 379–404. In: Pimentel, M., Fukf, H. (Org.). Sistemas Colaborativos. Rio de Janeiro—RJ: Elsevier-Campus-SBC. p. 416 (2011)

Franciscato, C.E., Martins, E., Jorge, T.M., Medeiros Neto, B.M., Martins, G.L., Werdemberg, A., ... Bueno, J.V.: Inovações no Jornalismo-Mesa Coordenada da Rede Jortec. In: 15° Encontro da SBPJor (2017)

Fuks, H., Raposo, A.B., Gerosa, M.A., Lucena, C.J.P.: Do modelo de colaboração 3c à engenharia de groupware. Simpósio Brasileiro de Sistemas Multimídia e Web–Webmidia 0–8 (2003)

Fuks, H., Raposo, A.B., Gerosa, M.A., Pimentel, M., Filippo, D., Lucena, C.D.: Teorias e modelos de colaboração. Sistemas colaborativos, pp. 16–33. Elsevier-Campus-SBC, Rio de Janeiro—RJ (2011)

García, R., Perdrix, F., Gil, R.: Ontological infrastructure for a semantic newspaper. In: Semantic Web Annotations for Multimedia Workshop, SWAMM (2006). Accessed on 22 May 2015. http://www.image.ntua.gr/swamm2006/resources/paper07.pdf

Ishikawa, E., Medeiros Neto, B.: Newsroom 3.1: incorporating social media management in semantic newsrooms with flexible business process. Conference: Seminário Hispano Brasileiro de Pesquisa em Informação, Documentação e Sociedade. Escola do Futuro ECA\USP—São Paulo, November, 11–14 (2019). Available at: http://seminariohispano-brasileiro.org.es/ocs/index.php/viishb/viiishbusp/paper/view/602

Larrondo, A., Domingo, D., Erdal, I.J., Masip, P., Van den Bulck, H.: Opportunities and limitations of newsroom convergence: a comparative study on European public service broadcasting organisations. Journalism Stud. **17**(3), 277–300 (2016)

Maia, K.B.F., Agnez, L.F.: A convergência digital na produção da notícia: Dois modelos de integração entre meio impresso e digital. Ponencia presentada en el Colóquio Internacional Mudanças Estruturais no Jornalismo–Mejor (2011). Brasília, jul. Recuperado de http://www.mejor.com.br/index.php/mejor2011/MEJOR/paper/view/73

Marcondes Filho, C.: Ser jornalista: o desafio das tecnologias eo fim das ilusões. Paulus (2009)

Medeiros Neto, B., Ishikawa, E.: Newsroom 3.1: incorporating social media management in semantic newsrooms with flexible business process. Conference: http://seminariohispano-brasileiro.org.es/ocs/index.php/viishb/viiishbusp/paper/view/602. At: Escola do Futuro ECA \USP—São Paulo—11 à 14 de Novembro de 2019 (2019)

Medeiros Neto, B., Ishikawa, E., Ghinea, G., Grønli, T.M.: Newsroom 3.0: managing technological and media convergence in contemporary newsrooms. In: Proceedings of the 52nd Hawaii International Conference on System Sciences (2019)

Molleman, E., Nauta, A., Jehn, K.A.: Person-job fit applied to teamwork: a multilevel approach. Small Group Res. 35(5), 515–539 (2004)

Laje, N.: Ideologia e técnica da notícia. rev. e ampl. Florianópolis: Insular (2012)

Oliveira, E.C., Ishikawa, E., Horinouchi, L.H., Granja, T.H., de A Nunes, M.V., Rodriguez, D., ... Ghinea, G.: Designing an ontology based Zika virus news authoring environment for the semantic web. In: Proceedings of the 8th International Conference on Management of Digital EcoSystems, pp. 197–203. ACM (2016

Pimentel, M., Gerosa, M.A., Filippo, D., Raposo, A., Fuks, H., Lucena, C.J.P.D.: Modelo 3C de Colaboração para o desenvolvimento de Sistemas Colaborativos. Anais do III Simpósio Brasileiro de Sistemas Colaborativos 58–67 (2006)

Quintarelli, E., Resmini, A., Rosati, L.: Information architecture: facetag: integrating bottom-up and top-down classification in a social tagging system. Bull. Assoc. Inf. Sci. Technol. 33(5), 10–15 (2007)

Rublescki, A., Barichello, E.: Jornalismo colaborativo e redes sociais no mainstream: estudo comparado do jornal zerohora. com e do washingtonpost. com. Rumores, 7(14), 99–118 (2013)

Salaverría, R., García-Avilés, J.A., Masip, P.: Concepto de convergencia periodística. In: García, X.L., Fariña, X.P., (coords.) Convergencia digital. Reconfiguración de los medios de comunicación en España. Santiago de Compostela: Servicio editorial de la Universidade de Santiago de Compostela, pp. 41–64 (2010)

Salaverría, R.: Los labs como fórmula de innovación en los medios. El profesional de la información 24(4) (2015)

Schauer, B., Zeiller, M.: E-collaboration systems: how collaborative they really are. In: Proceedings of COLLA 2011—The First International Conference on Advanced Collaborative Networks, Systems and Applications (2011)

Undurraga, T.: Making news, making the economy: Technological changes and financial pressures in Brazil. Cultural Sociol. 11(1), 77–96 (2017)

Wazlawick, R.S.: Metodologia de Pesquisa para Ciência da Computação Elsevier Editora. São Paulo (2009)

Zaragoza-Fuster, M.T., García-Avilés, J.A.: The role of innovation labs in advancing the relevance of Public Service Media: the cases of BBC News Labs and RTVE Lab. Commun. Soc. 33(1), 45–61 (2020)

Gheorghita Ghinea is a Professor in Mulsemedia Computing in the Department of Computer Science, at Brunel University. Dr. Ghinea's research activities lie at the confluence of Computer Science, Media and Psychology. In particular, his work focuses on the area of perceptual multimedia quality and how one builds end-to-end communication systems incorporating user perceptual requirements. Dr. Ghinea has applied his expertise in areas such as eye-tracking, telemedicine, multi-modal interaction, and ubiquitous and mobile computing, leading a team of 8 researchers in these areas. He has over 300 publications in his research field and is the Editor in Chief of the International Journal of Pervasive Computing and Communications. Currently, his research pursuits are centered on extending the notion of multimedia with that of mulsemedia a term which he has put forward to denote multiple sensorial media, ie. media applications which engage three or more of the human sense. His work has been funded by both national and international funding agencies and has been covered by the BBC, Telegraph, and Forbes magazine, among others. He consults regularly for both public and private institutions in his areas of expertise.

Benedito Medeiros Neto Post-Doctorate/Informatics: Semantic Framework for Journalism by CIC/IE/UnB (2018). Post-Doctorate: Digital Literacy and Mobile Learning by the School of Communication and Art/USP (2014). PhD in Information Science: Evaluation of Digital Inclusion programs, by FCI/UnB (2012). Master in Operational Research/Graph Theory by EST/UnB (1981). Specialist in Electrical Engineering/Artificial Intelligence by UnB (1986). Electrical/Telecommunications Engineer by UnB (1975). Visiting Professor at Computer Science Department, Brunel University, London/UK, May 2018. Project Scholar/MEC/MCTI/CAPES/CNPq/FAPs No. 09/2014. Researcher and Professor at UnB/IE/CIC and FAC/UnB. Associate Researcher at Escola do Futuro\USP (2014-). Consultant/Evaluator of FAPESB/BA. Reviewer at IGI Global. Associate of ASSOCIAÇÃO PROFISSÃO JOURNALISTA (2019). Participant of the GT01/ENANCIB; SIMEDUC/UNIT/Aracaju; Ibero-American Magazine of CI/Faculty of Information Science/UnB. PROFESSIONAL LIFE: Director of Innovation and Development at IBrTec (2019-); At Ministry of Communications: Consultant for Digital Inclusion; Coordinator of Knowledge Management and Evaluation of the GESAC Program (2012). At ECT he was Director Manager (2002), Advisor to the Vice Presidency (1999), Advisor/Technical Support (FAT) to the Directorate of Technology and Infrastructure (1998) and Senior System Analyst (2007). He was Head of Telecommunications Section of the Telebrás System (1978). He was Professor at ESAP/ECT (1988), Professor at CEUB/Brasília. DEVELOPMENT AND RESEARCH AREAS: Computer Science, Information and Communication; Network Engineering; ICT teaching; Informatics and Society; Collaborative Systems and Web; Semantic Web; Digital Inclusion; Digital Cities; Competence in Information, Social Networks and Evaluation of Innovation Programs. CNPq RESEARCH GROUPS: (a) JorTec/SBPJOR; (b) Journalism and Memory in Communication; (c) Technology and Digital Narratives; (d) Competence in Information.

Maria de Fátima Brandão Ramos PhD in Social, Work and Organizational Psychology from the University of Brasília (2009), Master in Computer Science from the Federal University of Rio Grande do Sul (1984), Specialist in Organizational Coach from Newfield Consulting and UCB (2001) and graduated in Data Processing by UnB (1978). She is a founding professor of the Department of Computer Science at the University of Brasília, where she has been working since 1983. She coordinated the implementation of the Bachelor's Degree in Computer Science and the first Licensed Teacher's Degree in Computer Science in Brazil. Participated in the Evaluation of the PROINFO Program (MEC /SEED) and in the implementation of the National Higher Education Evaluation System (SINAES) at INEP. She coordinated the evaluation of the Casa Brazil Project, a federal government action for the digital and social inclusion of MCT/SECIS. Coordinates the Extension Network Program for Digital Inclusion—REID, started in 2010 at UnB. Worked in the production and offering of disciplines for UAB/UnB and as a teacher in the Professional Master's program in Sustainable Management of Lands and Indigenous Peoples at the Center for Sustainable Development and in the Professional Master's in Information Security at the Department of Computer Science at UnB. She served as Technical Director of Graduation of the Dean of Undergraduate Education and Institutional Prosecutor of UnB from 2013 to 2016. She is the Technical Coordinator of the MDM Research Project—Multimodal Digital Media under the International Cooperation Program CAPES—PVE Subprogram 2014.

Edison Ishikawa He is a professor of the Department of Computer Science of the Brasília University (UnB) since 2014. He received the PhD degree in Systems and Computer Engineering from the Federal University of Rio de Janeiro (COPPE/UFRJ) in 2003, the M.Sc. degree in Informatics from Pontifical Catholic University of Rio de Janeiro (PUC-RIO), the B.E. Degree in

Computer Engineering from Military Institute of Engineering and the B.S. Degree from Agulhas Negras Military Academy. His current research interests focus on Semantic Computing, Systems of Systems Engineering, Distributed Systems and Security. At UnB he participates in research projects in those areas, with funding from Brazilian government agencies and institutions such as CNPq, FAP-DF and Brazilian Army. Participated in the implementation activities of the MDM Project: A model proposal for a semantic framework of a collaborative environment for information management in journalistic writing.

Conversational Competencies in Convergent Newsroom Environments

Ana Cristina Carneiro dos Santos, Gentil José de Lucena Filho, Gheorghita Ghinea, Lillian Maria Araújo de Rezende Alvares, Ébida Rosa dos Santos, and Maria de Fátima Ramos Brandão

Abstract The research investigates the importance of the conversations and their potential to contribute to the news production routines in journalistic newsroom. Based on the Ontology of Language, this study was initially characterized as a descriptive, applied and exploratory research. By developing the instrument of data analysis, called Matrix of Senses, the research started to be considered also methodological, based on Grounded Theory Methodology (GTM). One of the important contributions of this work is to situate the importance of the conversations, giving them formal, theoretical, philosophical and methodological grounded visibility for the news production process. The developed *Matrix of Senses* contributed to the explanation of patterns of behavior and presents itself as an instrument that can be customized and replicated to other contexts in which the conversations play a relevant role.

Keywords Ontology of language · Newsrooms · Conversational competencies

A. C. C. dos Santos (✉) · L. M. A. de Rezende Alvares · É. R. dos Santos ·
M. de Fátima Ramos Brandão
University of Brasilia, Brasilia, Brazil
e-mail: ana.carneiro@unb.br

L. M. A. de Rezende Alvares
e-mail: lillianalvares@unb.br

M. de Fátima Ramos Brandão
e-mail: fatimabrandao@unb.br

G. J. de Lucena Filho
Conversational Intelligence Laboratory, Brasilia, Brazil

G. Ghinea
Brunel University, London, UK
e-mail: george.ghinea@brunel.ac.uk

B. Medeiros Neto et al. (eds.), *Digital Convergence in Contemporary Newsrooms*,
Studies in Systems, Decision and Control 370,
https://doi.org/10.1007/978-3-030-74428-1_6

87

1 Introduction

Communication processes in newsroom environments involve ontological aspects that are determining of their success. Based on Heidegger's philosophy, the term ontology referred to in this article considers the possibilities of "being" in its historical perspective and in its creative potential. More specifically, it refers to the universe of the Ontology of Language, which studies the nature of human "being" and its new forms of coexistence.

This article is part of an exploratory study that investigates conversations, and considers their universal and general characteristics. It is universal because everywhere in the world, regardless of language, people talk. And it is general, because actions such as requests, offers, and promises, even in different locutions, creeds, ethnicities, or social classes, in an illocutionary manner, have the same meaning (Searle 1980).

Based on these characteristics, this article investigates the importance of the conversations, in particular, conversations for coordination of actions and realization of commitments related to the process of news production in journalistic newsroom environments. Its relevance lies in the potential of contributing to the efficiency and effectiveness of news production routines, deeply modified and vulnerable by the new forms of configuration of the collaboration networks and their new information and communications technologies (ICTs), more specifically, their new media and supports of multimodal digital communication.

2 Theoretical Foundation

Influenced by the contributions of Nietzsche, Wittgenstein, Heidegger, Maturana, and Flores, the Chilean sociologist and philosopher Rafael Echeverría, proposes a specific and particular articulation of the contributions of these authors and, supported by them, offers a new integrative conception of the human phenomenon, called Ontology of Language.

The Ontology of Language has three postulates:

(i) the interpretation of human beings as linguistic beings;
(ii) the interpretation of language as generative; and
(iii) the understanding that human beings develop in and through language (Echeverría 1997).

To support the development of this article, some approaches and thematic relations between concepts relevant to the understanding of the meaning of effective conversations within the scope of Language Ontology are described below.

2.1 Organizations Such as Commitment Networks

In the 1980s, Winograd and Flores (1998) already worked with the idea, later reinforced by authors such as Echeverría (1997, 1998, 2002), of which organizations exist as networks of directive commitments. Directives (which include requests, offers, and queries) and commitments (which include promises, acceptances, and rejections) are responsible for accomplishing the tasks and thus for achieving individual and organizational results. Such directives and commitments are made up of "acts of speech," a theory initially elaborated by John Langshaw Austin (1911–1960) and later developed by Searle (1980), who refers to language as a form of action.

For Austin (1962), "every saying is a doing" and is also "a way to interact with the world". Locutionary (act of pronouncing a statement; what is said), illocutionary (acts made through the pronunciation of a statement; actions embedded in what is said) or perlocutionary (effects produced), speech acts, according to Winograd and Flores (1998), correspond to the core of the entire work process carried out in the organizations through conversations.

2.2 Conversational Commitments

According to Kofman (2002, v. II), individuals or teams demonstrate their ability to fulfill commitments in the effectiveness with which they perform their tasks, establish the necessary relationships of trust to fulfill them and invest in ensuring the identity of individuals, groups or organizations involved in these relationships and tasks. According to the author, conversational commitments affect three levels: Task (T), Relationship (R) and Identity (I). Then, based on Kofman (2002, v. II) and Lucena Filho (2012), we provide a description of each:

- T: The goal is to coordinate actions to get the desired results. It refers to the task with which the individual or group commits to perform. It is concerned with the quality of the results of this execution.
- R: The objective is to generate bonds of trust, in order to enable effective coordination of future actions. It is concerned with the quality of relationships between people.
- I: The goal is to act with integrity and dignity. It is concerned with the coherence between the commitments, intentions and actions of an individual and with their dignity.

From the perspective of Ontology of Language, the conversational commitments correspond to a coordination of actions, where all commitment is a structure (TRI) and is also the key condition for the results, individual and organizational, to be achieved (Kofman 2002, v. II). For Kofman (2002, v. II), "the ability to receive and make commitments is one of the characteristics that define people".

In addition to the basic and involved speech acts and the TRI triad, other factors are necessary for the effectiveness of conversations where commitments are structured, assumed and executed. Among these factors, trust is an essential part of this dynamic (Echeverría 1998).

According to Echeverría (2002) and Flores (1994), the construction of trust is carried out from three domains: sincerity, competence and responsibility. By making a promise, people commit themselves in these three domains (Echeverría 2002). Our performance and public judgment of our performance in each of these domains is one of the basic ways in which we construct our public identity (Flores 1994).

Promises are linguistic actions by excellence of the coordination of actions between individuals. They correspond to a complex phenomenon that includes a number of linked actions called "promise cycle" or "action coordination cycle" (Echeverría 1998). By exploring basic elements (steps) of the action coordination cycle, Flores (2015), states that in making a promise, we are telling the other person that we are able to fulfill the conditions set out for the commitment.

2.3   Effective Conversations

According to Echeverría (1997), in its basic nucleus, the Ontology of Language focuses on the interpretation of three terms: human beings (they are linguistic beings), language (is generative) and action (generates the being). Based on this postulate, and understanding that organizations perform their actions through the conversations, Fig. 1 below presents the basic elements (steps)—composed of speech acts—that constitute the "conversations for actions" present in any organization (Flores 1994, 1996, 2015; Echeverría 1998, 2002).

As shown in section "Conversational Commitments", it is the trust links that enable effective coordination of future actions (Kofman 2002, v. II). Moreover, trust links lie at the center of the cycle of a promise that trust and the shared space of concern (Echeverría 1998). Without the presence of these elements—trust and shared space of concerns—it is not possible to coordinate actions effectively. Effectiveness, in the context of this article, consists in achieving efficacy (which achieves the proposed objective) with efficiency (that is done with excellence) and sustainability (with conditions/capacity of continuity). It is worth emphasizing that the shared space of concerns exists within a culture and that this culture varies from organization to organization. So what is obvious to one group may not be obvious to another. Hence the need to take care of networks of organizational commitments so that speech acts (requests, offers, promises and declarations), used in a given context, make sense for that community that is there, with regard to the actions embedded in what is said.

Based on the studies presented so far, this article assumes the understanding that effective conversations are those that, when coordinating actions (T) to obtain the desired results, take care of the relationships (R) between the interlocutors and the

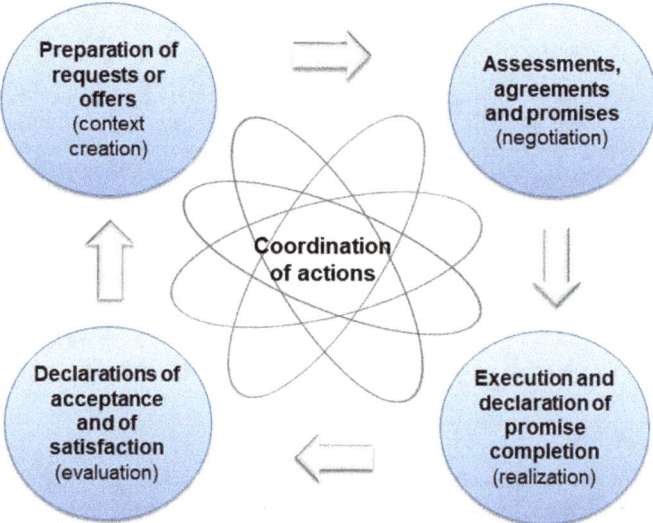

Fig. 1 Basic elements (steps) of a "Conversation for Action". *Source* Elaborated on the basis of Flores (1994, 1996, 2015) and Echeverría (1998, 2002)

identity (I) of the actors involved with the conversation. Thus, a conversation can be understood as effective when the three levels (T, R, and I) constitutive of the commitment assumed are simultaneously considered and satisfied.

On the other hand, another possible characterization of effective conversations is in the accomplishment of the steps of the cycles of coordination of actions, carried out in the organizational environments. Initially developed by Flores (1994) and later developed by Echeverría (1998), Kofman (2002, v. II), Lucena Filho (2010) and others authors, the cycle of coordination of actions—also called the cycle of promises or steps of a conversation for action—shows a relevant construct to verify the effectiveness of the conversations in the organizations.

2.4 Mental Models

Considering that speech acts present in the conversations correspond to the core of the entire work process carried out in the organizations and that the main role of the manager is to take care of the articulation and activation of commitments within the organizational network in order to achieve its results, the leaders must be aware of the efforts required to generate and maintain effective conversational networks (Winograd and Flores 1998; Flores 2015).

However, according to Kofman (2002, v. I, p. 263), the great challenge of the objective theory of communication is not to fall into the trap of thinking that "what I

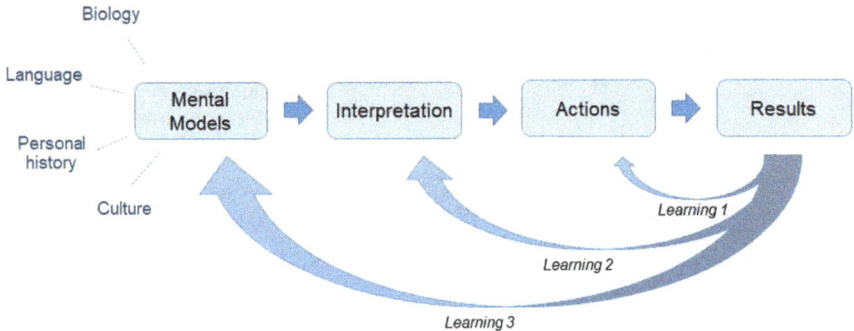

Fig. 2 Map of the observer, action and result. *Source* Elaborated on the basis of Kofman (2002, v. I) and Echeverría (2009)

say is what the other listens to" and "what I listen to is what the other says". For the author, "what each individual hears is conditioned by their mental models, because between what one says and what the other listens to, there are a series of filters that can generate large gaps between the sense of the said and the sense of the heard" (Kofman 2002, v. I, p. 264).

As shown in Fig. 2, there are four sources that determine the "usual" response we give to certain circumstances programmed in our mental model: biology, language, personal history and culture (Kofman 2002, v. I; Echeverría 2009).

From the "Map of the Observer, Action and Result", elaborated on the basis of Kofman (2002, v. I) and Echeverría (2009), Lucena Filho (2014) states that conversations, as instruments of human performance, act on three levels: actions, interpretations and mental models. For each of these levels there is a corresponding level of learning (and complexity), i.e., first, second, and third orders of learning.

Still according to Lucena Filho (2014, p. 6), "ultimately, conversations aim to transform the way of being (mental models and interpretations) of the Observer. For this, they use techniques and conversational tools. Their effectiveness presupposes an ethical relationship based on respect for: legitimacy, autonomy, freedom and difference of the other."

2.5 The Effective Speaking and Listening

Human communication has two facets, "speaking" and "listening." Effective speaking is only possible if it is followed by effective listening, because it is the effective listening that gives meaning to what is said. Listening is the fundamental factor of language: "We speak to be heard" (Echeverría 1997, p. 81).

A contribution that is in line with Lucena Filho (2010, 2014) on the use of conversational techniques and tools is found in Kofman (2002, v. II), who talks about conversations from two angles: exposing and inquiring. According to the author, inquiring is not usually present in the conversations with the same frequency that the exposing. People are not always open to "listening" to what the other person has to say. And in "talking", it is customary to observe attitudes of imposition of ideas or opinions, just as it is also customary to perceive a certain competitiveness in much of the conversations in organizations. Kofman (2002, v. II) calls these phenomena unproductive exposing and unproductive inquiring.

In contrast, productive exposing "is a way of opening our reasonings to others, to help them understand why we think what we think" (Kofman 2002, v. II, p. 98). The productive inquiring "is a way of discovering the reasonings of others and helping them to expose not only what they think but also why they think what they think." (Kofman 2002, v. II, p. 102).

With regard to effective listening, the challenge is even greater, since it is she who directs the entire communication process. Effective listening lies in the ability to perceive illocutionary speech acts present in the locutionary speech acts. For this, certain distinctions are necessary. According to Maturana and Varela (2001), the phenomenon of communication does not depend on what is delivered, but on what happens with what is received. Therefore, the decisive factor in the equation of listening is interpretation. The understanding of the sources that determine the mental models of the observers is one of the ways that allow to provide distinctions that contribute to the process of interpretation.

Listening, exposing, and inquiring are part of the universe of conversational competencies. By enabling people to consciously use words to articulate commitments and invoke effective coordination of their actions, it is possible to reduce misunderstandings and mistakes that prevent many organizations from realizing their potential (Flores 2015).

In this context, considering that basic talents and experiences contribute to new forms of "being" and "doing", this study adds a fifth level (experience) to the four levels (knowledge, skills, attitudes and values) in which, according to Lucena Filho (2010), conversational competencies manifest themselves. In this way, complementing Lucena Filho (2010), this article understands conversational competence as the capacity of, by conversing, to mobilize, articulate and put into action, in a sustainable way, knowledge, skills, attitudes, values and experiences necessary for the efficient and effective performance of activities required in work and life, in general.

3 Conversational Distinctions Applied in the Context of Newsrooms

Although we see with our eyes, we observe with our distinctions. According to Echeverría (1997), our ability to establish differentiations from certain particularities, that is, our distinctions, are our own constructions and make us different observers. People with different sets of distinctions live in different worlds. Our traditions of distinctions are different. Therefore, a distinction only makes sense in the context of a certain tradition of distinctions.

3.1 Journalistic Newsrooms

Over the past few years journalistic newsrooms environments have shown a growing concern, not only with the quality of information production in the various supports and ICTs, but above all with the digital convergence of the multimodal production process and its effective dissemination in the their own time, considering the needs and specificities of the various user information profiles.

Jenkins (2009) uses the term media convergence, to explain the movement to which the media have joined to adapt to the internet and thus distribute their products. The main drivers of convergence are technological changes and the digitization of processes. In this way, convergence works as a response to the new communication processes that emerged with Information and Communication Technologies (ICT) and starts to affect several dimensions of information production (Silva et al. 2013, p. 51), including the form how users and professionals involved in the news production process coordinate actions among themselves.

The multifaceted context of convergence, however, prevents a unanimous definition, as advocated by Avilés et al. (2014) when revisiting the concept. For the authors, convergence appears as a phenomenon that influences the media system, shaping different dimensions of communication. The convergence of newsrooms, although it is an ongoing process, has caused changes in the news production process. As the changes impact the workflow of professionals, such impacts are also reflected in the processes of coordination of actions, in the relationships between the interlocutors and in the identity of the actors involved in the conversation, affecting the triad T-R-I. For example, the prioritization of online content by printed newspapers is one of the verified movements (Lenzi 2018) that act directly on the triad. The newsroom, which used to be regulated at a slower pace, is now being pressured into a cycle of innovation in newspaper companies, in which the speed of content publication is also a determining element. Therefore, if we consider that speech acts are the central nucleus of the productive process of journalism and that it is necessary to articulate and activate commitments within the newsroom for the newspaper to function, automatically there is a greater need for attention and efforts so that effective conversational networks are created and maintained in these environments in order to obtain better results or satisfactory results.

In this regard, Lenzi (2018) demonstrates that within the setting of convergence in the contemporary international context of newsrooms, the practice of collecting and formatting journalistic content done by the same integrated professional group and the packaging of data according to each platform has been reinforced, while the editing has been carried out by different teams. Thus, once again the need for coordinated and conversed actions is demonstrated.

Additionally, the work and production processes increasingly incorporate the participation of information consumers and the demand for new products, which has modified the workflows and the networks of commitments, in their qualitative, quantitative and logistic aspects. Having overcome the idea of a merely receptive reader, today we are witnessing the content producing reader, the participatory audience (Tárcia 2007). The interaction on the part of the public ends up impacting the professionals' routine, requiring both technological and editorial policy strategies to deal with demands, suggestions and interactive routine. Editorial policies are a set of central rules of the journalistic organization, defining standards and behaviors of professionals in relation to the different aspects of production, being characterized as a guide that serves as a model for leaders or in the case of newsrooms for editors, to assist in the coordination of the teams and in reaching the final result which is the newspaper. However, with the rapid changes and the still unknown scenario, the agreements and internal conversations gain even greater relevance, being decisive for meeting the needs imposed by new media and maintaining the production process of the newsroom.

Social media platforms, for example, have been progressively adopted in newsrooms over the years and online communities have become an integral part of the news production process. As Heravi and Harrower point out, "Journalists monitor social media for news and content, use them to find sources and eyewitnesses, and take advantage of its wide reach to gain varied perspectives on interesting events" (2016, p. 1195). An example of networks that generate, reverberate and impact journalistic content is Twitter. As Lasorsa et al. (2012, p. 22) exemplify "Social network sites such as Twitter have helped the audience to become active in the news-creation process [...], where messages move back and forth and where users have the chance to interact with information".

Aligned with this process of change, some aspects observed in the data from the interviews carried out by the team of the Multimodal Digital Media (MDM) Project, presented below, were considered of special relevance for this study. Among them, the changes related to the insertion of the consumers of the news in the chains of commitments responsible for the production of the news—previously composed only by professionals inserted in journalistic newsroom environments, the frequent changes generated by new ICTs and, in particular, the possibility to agree and make commitments with people from different places and cultures.

3.2   Effective Conversations in Newsrooms

Based on the distinctions presented in the previous sections, effective conversations in journalistic newsrooms environments are those that, when handling the task coordination (T) to obtain the desired results, also take into account the relationships (R) of trust and the identities (I) of the people involved in the news production process. Moreover, given that journalistic newsrooms are networks of organizational commitments and, as such, constitute and fulfill promises to achieve their objectives, the effectiveness of the coordination cycles of actions carried out in newsrooms can provide significant information about the effectiveness of the conversations developed by their teams.

It is worth noting that, when using the TRI triad and the coordination cycle of actions as theoretical constructs for analysis of the secondary data presented below, distinctions are used on mental models, effective speaking and listening, trust, shared space of restlessness and other theoretical foundations applied to journalistic newsroom environments.

To help illustrate the narrative about the effectiveness of conversations in journalistic newsroom settings, Fig. 3 presents the steps of a conversation for coordination of actions (Fig. 2), where TRI levels of commitments are worked out from judgments based on trust between the parties, in order to satisfy certain future conditions—in the case of journalistic newsrooms, the production of qualified and timely news.

As illustrated, each step of the action coordination cycle deals with tasks (T) that are developed, through relationships (R), within the identity (I) of those involved, therefore, in the TRI triad.

The figure draws attention to the existence of several conversations within a larger conversation or several commitments within a broader commitment. The result of each "small" commitment affects the other commitments agreed in the

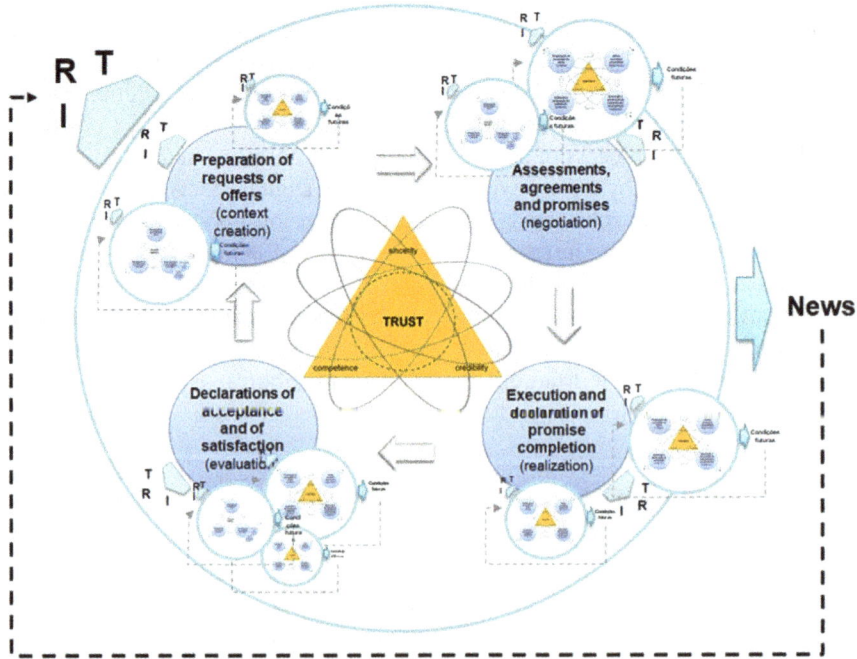

Fig. 3 Cycle of coordination of actions in newsroom environments. *Source* Elaborated on the basis of Flores (1994, 1996, 2015), Echeverría (1997, 1998, 2002) and Kofman (2002, v. II)

network of conversations, where the actions are being coordinated, therefore affecting the "greater" commitment in which everyone is involved.

In newspaper editorial environments this is easily observed when, for example, the publication of a single news item depends on several articulations of actions between different teams of photography, writing and making content available in multiple media. All this happens at the same time and with a strong dependence between the delivery of a commitment and the beginning, continuation and/or completion of another.

This study considers that the TRI levels of the commitments assumed by the networks present in newsrooms generate results and are impacted by them, providing their own evolution. In this way, the result is interaction as a whole, where in producing news (task), the network reconfigures itself (relationships) and the identities (integrity and dignity) of the professionals involved in the commitment also change. TRI levels are absolutely complementary and interdependent; that is, they need to be congruent with one another.

3.3 Primary Data

Between April 2015 and May 2017, researchers from the MDM Project conducted a series of visits to the editorial offices of the following institutions: BBC (England), *Correio Braziliense* (Brazil), *La Nación* (Costa Rica), *O Globo* (Brazil) and Reuters (England). During these visits, editors, sub-editors, reporters, editors and representatives of the areas of photography, design and ICTs of the visited newspapers were interviewed.

As presented by Oliveira (2017), the visits included the participation of staff meetings; observation of the work dynamics of newspapers; knowledge of the processes of receiving and ascertaining facts; survey of part of the software used, etc. The primary data referred to in this article correspond to the interviews recorded. The secondary data correspond to the excerpts extracted from the primary data.

Inserted in the field of Information Science, this article seeks to explore the theoretical aspects related to conversational competencies in the news production process, presented in the previous sections. For this, in addition to researched references about effective conversations, from the perspective of Language Ontology, excerpts were extracted from the recordings of the interviews that dealt with some level of coordination of actions among the newsroom teams visited. Subsequent to this selection of primary data and extraction of secondary data, from the theoretical foundations presented, data analysis, described in section "Analysis and Results", was performed.

As the data revealed, news production processes are fundamentally conversational processes. Moreover, each environment works according to its culture. News generated in different newsrooms may be similar, but the networks of commitments that generate them behave in different ways. Hence the relevance of an analytical tool capable of providing subsidies for expanding the conversational competencies of the commitment networks involved in the news production process.

4 Methods and Procedures

Based on the Ontology of Language—which understands that behaviors are significantly impacted by language, initially this study was characterized as a descriptive, applied and exploratory research. However, after the application of the first instrument of analysis, by focusing the data and field of study and developing a second instrument of data analysis, the research was also considered methodological, based on the Grounded Theory Methodology (GTM).

Figure 4, summarizes the main challenges and lessons learned during the process of extracting, analyzing and treating the data originally available for the development of the research.

Developed by Glaser and Strauss, GTM uses procedures to inductively develop a theory derived from the data. Its basic premise is that "everything is information" (Brown 2013). According to Gasque (2007, p. 90), in the GTM, "the researcher builds a theory from the specific observation of the phenomenon". According to the author, this involves a theoretical sampling process (strategy of gradual definition of the sample, theoretical sensitivity of the researcher and control by the theory in formation), codification (conceptualization of data, comparison between phenomena, cases and concepts and foundation of a theory) and writing theory (relations between categories, presentation of findings and description of theory).

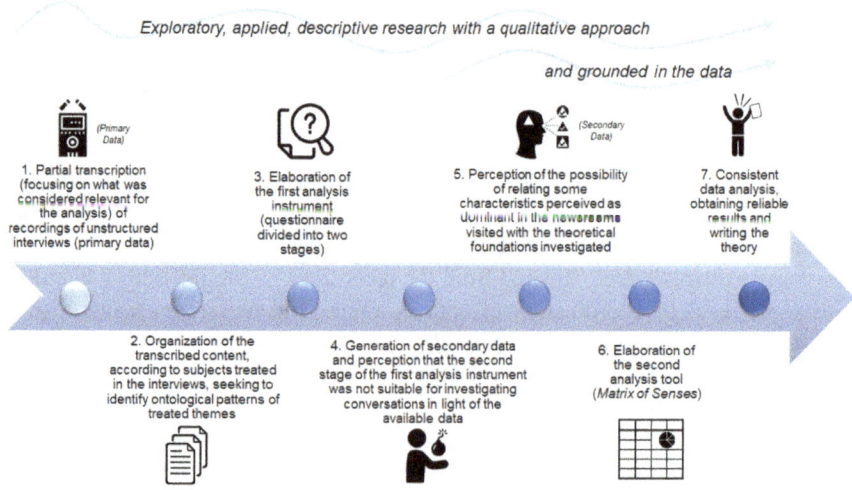

Fig. 4 Methodological trajectory. *Source* Lucena Filho and Carneiro (2019). Free translation

4.1 First Instrument of Analysis

Briefly, the first instrument of analysis was divided into two stages. Stage I referred to the selection of excerpts (secondary data) from the recordings of the interviews (primary data) that presented some level of coordination of actions and accomplishment of commitments by the teams responsible for the production of the news. Stage II corresponded to a questionnaire, answered based on the excerpts (secondary data) from stage I, with generic and specific questions about the theoretical foundations presented here, based on the work of Ávila et al. (2017) which, in turn, was based on Andrade (2009), Santos (2008), and Braga (2007).

In this phase II, the generic questions, distributed in six items, sought to raise aspects related to effective speaking and listening, trust, sharing of concerns and other related points in the theoretical foundations. The specific questions were based on the TRI triad, the coordination cycle of actions and the concept of effective conversations in newsrooms environments presented.

4.2 Second Instrument of Analysis

The second instrument of analysis, Matrix of Senses, presented in section "Analysis and Results", proposes a relation, according to the authors' interpretation, between the dominant characteristics present in the news production process, of the newsrooms visited (characteristics observed in the secondary data from stage I of the first instrument of analysis) and the variables of the theoretical foundations investigated.

Both the list of generic and specific questions present in stage II of the first instrument of analysis as well as the structure of the Matrix of Senses (which corresponds to the second instrument of analysis applied in this study) represent scripts that can be used to analyze and guide the conversations in newsroom meetings, seeking its effectiveness (efficacy, efficiency and sustainability).

5 Analysis and Results

As discussed in section "Conversational Distinctions Applied in the Context of Newsrooms", this work observed the recordings of unstructured interviews conducted by researchers of the MDM Project with journalism professionals from the five (5) newsrooms visited. As the interviews dealt with diverse issues of interest to the MDM Project and, therefore, did not address questions directly related to effective conversations in newsrooms environments, the application of the first analysis instrument was preceded by the following tasks:

(a) Partial transcription (with focus on what is relevant to the analysis) of recordings of unstructured interviews (primary data) in text or slide format.

(b) Attempt to organize the transcribed content, according to the subjects discussed during the interviews, seeking to identify patterns of subjects.

After performing these tasks, the first analysis instrument was then applied. This application was carried out by the researchers responsible for this article, through the completion of the two stages of the instrument. Accordingly, stage I used the primary data (recordings of the interviews) to generate the secondary data (excerpts that presented some level of coordination of actions and realization of commitments related to news production), while stage II (later discarded) used the excerpts (selected in stage I) to answer their questionnaire and generate graphs and analysis from the answers obtained (not used in this study).

The use of primary data, collected in a previous study to the study presented here, raised by observers different from the authors of this article and with different concerns from those explored here, did not allow us to obtain useful answers to the questionnaire that comprised stage II of the first instrument of analysis. In this way, it was realized that it would not be possible to investigate with some level of depth the ongoing conversations in the newsrooms in the light of interviews recorded by the MDM Project team. These primary data demonstrated that there was no singularity, script or structuring agenda that revealed a certain degree of homogeneity in the recordings. In addition, no evidence was found that there was a method behind the interviews. This explains the diversity of the objects found in the primary data.

However, the attempt to answer the generic and specific questions of stage II of the first instrument of analysis, based on the use of excerpts (secondary data from stage I) of the recordings of the interviews (primary data) that presented some level of coordination of actions and the accomplishment of commitments by the teams responsible for the production of the news, although not fulfilling the intended objective, demonstrated, according to the interpretation of the authors, the possibility of relating characteristics perceived as dominant in the essays visited with the theoretical foundations investigated.

Thus, based on the GTM, a second instrument was developed and applied, the "Matrix of Senses" presented below, which was based on the "affirmations of support"[1] (pronounced by the professionals interviewed in the newsrooms) extracted and interpreted from the secondary data (from stage I of the first analysis instrument), among them: "We have meetings every day. Before the general meeting with all the editors there is a meeting of each editor with his reporters. The emergence and popularization of social media generate profound impacts on

---

[1]The term "affirmations" in the Ontology of Language (Echeverría 1997, p. 42) denotes a "linguistic act" with its own characteristics; in particular, brings with it the imperative necessity that what is said may be accompanied by one or more factual evidence(s), unequivocally verifiable. In the case in focus, the highlighted excerpts from the interviews are referred to as "affirmations of support" because the interviews were recorded and can at any time be used as a support to evidence the manifestation of the "locutions" enunciated by the interlocutors participating in the interviews. And here it is important not to confuse "locutions" with "illocutions" (Echeverría 1997, p. 86).

journalistic practice and the development of new business models in the information industry." (*La Nación*); "It's always great to have coffee and talk because you end up building relationships. The public will often hear a story from another source and then come to the BBC to see if it's true. The international effort has many different aspects, among them, the translation.

Our editorial philosophy is that it is better to be late than to be wrong." (BBC); "With these multi-skills, people are taking better advantage of *N* media types. We have very good professionals from all different disciplines, they are in a moment of great transformation. People in one area are looking to learn about other areas. Skills need to make sense for both the professional and the organization. We try to get people to know each other, work together and cooperate." (Reuters); "In view of the changes in the journalist's profession, it is important that the reporter is multimedia and is increasingly prepared and updated on the different issues." (Correio). "The reporter does not receive the data passively and is not merely a mediator of information. He needs investigative, analytical and communicative skills." (*O Globo*).

The following matrix represents a process of ontological analysis of meaning and presents, in the author's hearing, a list of the main characteristics dominant and common to the essays visited, relating them to the main variables of the theoretical foundations investigated.

According to Maturana and Varela (2001), all organisms function because of their interaction with the environment, and the way this process occurs depends on the environment and the context in which they live. Thus, the listening capacity of the authors of this article, present in the conversational competencies of those who investigate the foundations presented here, helped to perceive what is predominant or not, in accordance with the context in which the information was extracted. In other words, it helped to make sense of the secondary data investigated (Table 1).

Based on Fig. 3, presented in section "Effective Conversations in Newsrooms", which synthesizes the conceptual frameworks worked on in this article, the analysis of the secondary data, demonstrated in the Matrix of Senses, showed that the process of news production in the context of all newsrooms visited happens through some level of coordination of actions, constant and with feedback, where each step —individual and organizational principles and drivers, preparation, execution and evaluation/ reflection—even if unstructured, involves care with: fulfilling the task (news publication), the relationship between actors (reporters, editors, photographers, designers, readers, spectators, competing newspapers and agencies, government, etc.), and the identity (dignity and integrity) of people reflected in the coherence between commitments, intentions and actions of the news professionals and newsrooms visited. Mental models, effective listening, exhibition capacity and productive inquiry, as well as aspects related to trust and openness to share concerns, were observed as variables to support the achievement of related steps.

The analysis of the secondary data allowed the authors to "listen" to what is not said in a locutionary way, but in an illocutionary way. Making it is possible to think that the newsroom teams visited do what they do, with the best of intentions and

Table 1 Matrix of senses

Key features in the news production process	Coordination cycle of actions (principles and guidelines, preparation, implementation and reflection)	Triad (task, relationship and identity)	Mental models (1st, 2nd and 3rd order learning)	Speak productive	Effective listening
Holding meetings to combine news bulletins. In some cases, more than once a day	• Staff meetings are examples of spaces for alignment oil the principles and guidelines related to the commitments that have been or will be undertaken, preparation of the activities that will account for the commitments, monitoring of what is performed and reflections on what has already been accomplished • Taking care of the steps that are part of an action coordination cycle contributes to meeting productivity and results	• The understanding of the task to be performed, the quality of the relationships between the participants and the alignment with the identity of those involved in the activities have a direct impact on the productivity and attainment of the commitments agreed upon at the meetings.	• Tariff meetings are learning spaces as they allow participants to assess and revisit their possibilities for action, interpretation and the mental models that guide them • This permission to learn direcdy impacts on the commitments agreed upon during the meetings as they encourage the development and creativity of the teams on the basis of positive results for the essays	• Tariff meetings are spaces for the exercise of exposure and productive inquiry • The courage and willingness of people to clarify and share their interpretations of behavior and results related to commitments contribute to the efficiency and effectiveness of the agreed actions	• Appropriate understanding of the step-by-step approach needed to coordinate actions, usually discussed at the staff meetings, which includes understanding the task to be performed, caring for the relationships and identities of those involved, depends directly on the listening meetings. Impact on the use of the time of meetings and the competences of the teams, and in obtaining results

(continued)

Table 1 (continued)

Key features in the news production process	Coordination cycle of actions (principles and guidelines, preparation. implementation and reflection)	Triad (task, relationship and identity)	Mental models (1st, 2nd and 3rd order learning)	Speak productive	Effective listening
The dependence of the actors (internal and external to newsroom environments) with different abilities to fulfill the tasks	• Care with the steps that are part of a cycle of coordination of actions contributes to the articulation of tasks, planning deadlines, costs and responsibilities, considering the different competencies of the actors (internal and external) involved in the activities	• Regardless of the location (internal or external to the essay), the means of communication (face-to-face or distance/virtual) in which the actors are working, the relationship level of a commitment must be taken care of so that the identities of the different professional profiles and the tasks are completed	• Mental models are formed by filters and interpretive structures in which different observers construct different realities • Recognizing the existence of these filters contributes to the understanding among the teams and, consequently, to the fulfillment of the challenges present in the newsroom, as the process of news production is carried out by different actors (observers) with their different biologies, languages, cultures and stories	• Exposure based on the speaker's concerns and the inquiry so that the other actor reveals his or her concerns that contributes to the understanding of what is said and, consequently, to the effectiveness of the commitments that involve different actors with different competencies.	• Commitments developed through partnerships require the ability to listen to the parties involved • Misrepresentations provoke misunderstandings between professionals w:ho depend on each other and, as a result, generate wear and tear, rework, waste time and money, and is frustrations and among other consequences

(continued)

Table 1 (continued)

Key features in the news production process	Coordination cycle of actions (principles and guidelines, preparation, implementation and reflection)	Triad (task, relationship and identity)	Mental models (1st, 2nd and 3rd order learning)	Speak productive	Effective listening
The need to present results daily	• Teams need to be able to commit and behave in a way that achieves daily results. Tor this, it is fundamental to know what should be done, by whom, when and how	• The daily production of news demonstrates the need to perform daily tasks by the teams	• Understanding the different levels of learning and the impact of mental models on the decisions and actions of the teams responsible for producing the news helps to make them more aware of their potentialities, difficulties and, above all, what leads them to this	• The daily production of news highlights the need for constant understanding of the tasks that must be delivered by the teams • These understandings come about through conversations that are negatively or positively impacted by the "talking" ability of teams	• The daily production of news highlights the need for constant understanding of the tasks that must be delivered by the teams • These understandings come about through conversations that are negatively or positively impacted by the ability to "listen" to teams
The need for constant dialogue among the actors to carry out the process of news production (search, determination, photography, editing and publication)	• Clarity regarding the guiding principles and directives, the preparation of activities, the monitoring of what is done and the possibility of reflections on the works, contributes to the effective dialogue	• The ability to take advantage of one's own competencies in an authentic way, respecting one's identity, contributes to the productivity of the relationship between the parties and to the execution	• Conversations transform the way of being (mental models with their filters and interpretive structures) of individuals (observers) • Considering that rt is through conversations that the	• The ability to make requests and offers, knowing the possibility of it being refused by another, as well as the freedom to inquire, trying to discover the reasonings of others; this contributes to the productive dialogue	• The openness to accept a refusal of a request not as a rejection or a lack of commitment, but as an action of sincerity and responsibility contributes to the productive dialogue and, consequently, to the articulation of

(continued)

Table 1 (continued)

Key features in the news production process	Coordination cycle of actions (principles and guidelines, preparation, implementation and reflection)	Triad (task, relationship and identity)	Mental models (1st, 2nd and 3rd order learning)	Speak productive	Effective listening
	between the teams responsible for producing the news • The existence of trust among team members is the basis for coordinating effective actions	of the tasks to be fulfilled	commitments of a team or organization are made, investments in the different levels of learning of the information professionals and, consequently, of the newsrooms themselves contribute to the efficiency, efficacy and effectiveness of the news production process	and, consequently, to the articulation of actions among those responsible for producing news.	actions among those responsible for producing news
The need for openness to the new, as ideas, information, criticism and solutions can come from anywhere and anytime—including the end users of information, including	• The accomplishment of commitments, through cycles of coordination of actions facilitates the reception, analysis, incorporation (or not) and feedbacks to ideas, criticisms and	• The openness and the desire to share information and ideas between professionals and users of the information help in the understanding and execution of the tasks	• Investments in 1st-, 2nd-, and 3rd-order learning contribute to broadening the perceptions of the teams • What impacts on autonomy, self-responsibility,	• The exhibition and the questioning of different points of view contribute in an effective way to the production of news • The courage and openness to the exhibition and	• Openness to the new in newsrooms, is directly related to the ability and availability of listening by those responsible for the news production process

(continued)

Table 1 (continued)

Key features in the news production process	Coordination cycle of actions (principles and guideline, preparation. implementation and reflection)	Triad (task, relationship and identity)	Mental models (1st, 2nd and 3rd order learning)	Speak productive	Effective listening
readers, listeners and viewers	solutions received from different places	related to news production	strengthening of identity and other aspects that favor openness to interact with ideas, information, criticism and proposals for solutions coming from different places and observers	inquiry of opinions, ideas, criticisms and solutions, including when it comes to divergent opinions, are necessary for the newsrooms that are truly committed to new ways of configuring the collaboration networks and their new ICTs	• Understanding that differences of opinion can enrich work is part of the exercise of effective listening
The need for constant improvement by information professionals (especially in technologies) and motivation to work and invest in rapidly changing, highly competitive environments that depend on mutual	• Clarity regarding the guiding principles and directives, the preparation of activities, the monitoring of what is done and the possibility of reflections on the work, contribute to explain what is necessary in order to	• Respect for the identity (dignity and integrity) of information professionals has a direct impact on their development, their relationship with others and the ability to fulfill their assigned tasks	• Understanding the impact of mental models with their learning cycles—1st, 2nd and 3rd Order—provides inputs for the development of teams' development programs. Both at the operational level (action), at the level of reflection	• Exposure and productive inquiry contribute to the sharing of desires and concerns of professionals and organizations • This sharing fosters an understanding of the needs and possibilities for improvement of both	• As it is listening and not speaking that gives meaning to what is said, effective listening is fundamental to understanding the desires and concerns that promote the improvement and motivation of the teams

(continued)

Table 1 (continued)

Key features in the news production process	Coordination cycle of actions (principles and guidelines, preparation, implementation and reflection)	Triad (task, relationship and identity)	Mental models (1st, 2nd and 3rd order learning)	Speak productive	Effective listening
cooperation between people	obtain the results of the newsrooms and the types of skills desired for the teams • It also contributes to the best use of the competence available in the teams, the alignment between the professional goals of the people and the organization, as well as managing schedules e schedulings		(interpretation) or at the deeper level (filters and structures) that guide the interpretations and actions of individuals • This understanding supports not only the training programs of the newsrooms, but also the decisions of individual improvement of the team components	parties—both the newsrooms and their teams	

According to the authors' interpretation, there is a relationship between the dominant characteristics of the news production process in the visited newsrooms (observed in the secondary data), and the variables of the theoretical foundations investigated with some level of effectiveness, because that is what is possible, depending on the level of conversational competencies they have.

This, as presented in the Matrix of Senses, the development of certain distinctions would enable the teams to perceive actions performed in speech (illocutionary acts) present behind expressed words and sentences (locutionary acts). This analysis gives space for reflections on what could be done so that the teams can improve the level of effectiveness (perlocutionary) of the conversations aimed at articulation, execution and evaluation of the actions responsible for producing the news.

The Matrix of Senses contributes to the understanding that the development of new distinctions, related to the theoretical foundations investigated, would allow that which the teams commonly identify in the day to day (manifested perceptions), would begin to be identified from the distinctions that would allow deeper perceptions (beyond what is manifested) regarding the commitments made and the process of coordination of actions necessary to fulfill them.

Thus, the relationship between the dominant characteristics of the newsrooms visited and the variables of the investigated theoretical foundations, presented in the Sense Matrix, provides inputs ("findings") for the mapping of types of conversational competencies capable of making information professionals (reporters, editors, photographers, designers) who are able to develop effective conversations and obtain better results in the news production process, in view of the new forms of configuration of the collaboration networks and the new ICTs, more specifically, the new media and supports of multimodal digital communication with which these professionals relate.

6   Final Considerations

The main objective of this article was to analyze characteristics of the conversations in newsrooms environments, in particular, conversations to coordinate actions and fulfill commitments related to the news production process. As discussed in section "The Effective Speaking and Listening", conversational competence entails the capacity to, by conversing, to mobilize, articulate and put into action, in a sustainable way, the knowledge, skills, attitudes, values and experiences necessary for efficient and effective performance of activities. This ability (to converse) is what moves efforts to coordinate actions, even unstructured, for news production. Therefore, for newsrooms to achieve more effective results, it is necessary to invest in the way the teams converse. Which means the development of their conversational competencies.

One of the important contributions of this work is in situating the importance of the conversations, giving them formal, theoretical, philosophical and methodologically grounded visibility for the news production process. If newsrooms find conversations to make them more or less effective, it will be possible not only to give greater visibility to what happens and what involves the dynamics of their teams, but also, and perhaps mainly, to protagonize in a responsible manner the ethical aspect of social contribution by producing and placing in the world the news produced in our journalistic media. However, as has been seen, understanding the importance of conversations requires observers and differentiated actions, which are acquired through distinctions of our own that associate with the perceptions we have as human beings make us different observers.

The study of the theoretical foundations and the instruments of analysis, especially the Matrix of Senses, presented here allowed the observation of the conversations present in the actions developed by the newsroom teams. The ontological analysis of the main characteristics dominant and common to the essays visited, related to the main variables of the theoretical foundations, provided interesting contributions to the understanding of important aspects related to conversational competencies and helped to elucidate findings (inputs) useful for the promotion of changes in behaviors of individuals and newsroom teams visited. Once the conversational dynamics are observed, it is possible to intervene in this respect, working aspects related to the perceptions and distinctions of the individuals, and thus contribute to the efficiency and efficacy of the news production routines performed by these teams.

Elaborated based on the GTM that seeks to understand the phenomenon based on emerging data, not on pre-conceived data, the Matrix of Senses has shown that it is possible to make an evaluation of the relation between the observed phenomena and the ontological dimensions of the conversations, with respect to their levels of action. This instrument contributed to the explanation of patterns of behavior, related to the universe of conversational competencies, present in the news production process of the newsroom environments visited. From the analysis undertaken, this article proposes to expand this study and presents the Sense Matrix as an instrument that can be customized and replicated to other contexts in which the conversations play a relevant role.

In addition to the investigations, analyses and contributions presented here, this research also reveals the possibility of future unfolding of the studies on the various dimensions of the conversations, especially from the perspective of complexity theory, which will allow a deeper look at the dynamics in the structures (TRI), present in the coordination cycles of actions and in the ontological dimensions of the conversations that involve, besides the language, studies on corporality and emotionality.

Acknowledgements Special thanks are due to CAPES (Coordenação de Aperfeiçoamento de Pessoal de Nível Superior) for granting the scholarship during the "sandwich" doctoral scholarship in London; to the Universidade de Brasília (UnB), and to Brunel University, London, for the support provided for the development of this work. This publication is also part of the actions carried out under the MDM Project (see Appendix). The project had the financial support of the Science without Borders Program of CAPES/CNPq, Process: 88881.068354/2014-01; and CAPES, Process: 88887.144822/2017-00.

References

Andrade, E.: Conversas: o fator chave para o gerenciamento de projetos (Masters thesis). Universidade Católica de Brasília (UCB), Brasília, DF, Brasil. Available at: https://bdtd.ucb.br: 8443/jspui/handle/123456789/1566

Austin, J.L.: How to Do Things with Words. Boston: Harvard University Press. ISBN: 9780674411524 (1962)

Ávila, J.C., Lucena Filho, G.J., Figueiredo, R.M.: Competências Conversacionais para a Governança Corporativa. *iSys—Revista Brasileira de Sistemas de Informação* **10**, 85–110. ISSN: 1984-2902 (2017)

Avilés, J.A.G., Kaltenbrunner, A., Meier, K.: Media convergence revisited: lessons learned on newsroom integration in Austria, Germany and Spain. J. Pract n. **8**, 573–584 (2014)

Braga, S.O.: O coaching ontológico como instrumento de desenvolvimento de equipes de trabalho, Masters thesis. Universidade Católica de Brasília (UCB), Brasília, DF, Brazil. Retrieved at: https://bdtd.ucb.br:8443/jspui/handle/123456789/1525 (2007)

Brown, B.: A coragem de ser imperfeito. Editora Sextante, Rio de Janeiro. ISBN: 9788575429594 (2013)

Echeverría, R.: Ontologia del lenguaje. Dolmen Ediciones, ISBN: 9789562012263 (1997)

Echeverría, R.: El ciclo de la promesa: eslabón básico de coordinación de acciones. Centro de aprendizaje y enseñanza 1–26 (1998)

Echeverría, R.: Confiança viga mestra da empresa de futuro. Ethos Reflexão **7**, 1–26. https://www.ethos.org.br/wp-content/uploads/2013/02/Reflex%C3%A3o071.pdf (2002)

Echeverría, R.: El observador y su mundo, vol. 1. Juan Carlos Sáez Editor, Buenos Aires, Granica. ISBN: 9789563060447 (2009)

Flores, F.: Promesas, confianza e identidad publica, pp. 1–12. Redcom Chile S. A, Santiago de Chile (1994)

Flores, F.: Creando organizaciones para el futuro. Dolmen Ediciones, Santiago, Chile. ISBN: 9789562012171 (1996)

Flores, F.: Conversaciones para la acción. Inculcando una cultura de compromiso en nuestras relaciones de trabajo. Lemoine Editores, Bogotá. ISBN: 9781508651888 (2015)

Gasque, K.C.G.D.: Teoria Fundamentada: nova perspectiva à pesquisa exploratória. In: Mueller, S.P.M. (Org.) Métodos para a pesquisa em Ciência da Informação, pp. 83–118. Thesaurus, Brasília. ISBN: 9788570626547 (2007)

Heravi, B.R., Harrower, N.: Twitter journalism in Ireland: sourcing and trust in the age of social media. Inf. Commun. Soc. **19**, 1194–1213. ISSN: 1369-118X (2016)

Jenkins, H.: Cultura da convergência, 2nd edn. Aleph, São Paulo (2009)

Kofman, F.: Metamanagement: a nova consciência dos negócios, vols. 1–3. Antakarana Cultura, Arte e Ciência/Willis Harman House, São Paulo. ISBN: 9788588262027, 9788588262034 and 9788588262041 (2002)

Lasorsa, D.L., Lewis, S.C., Holton, A.: E. Normalizing twitter. Journalism Stud. **13**(1), 19–36 (2012). https://doi.org/10.1080/1461670X.2011.571825

Lenzi, A.: *Inversão de papel*: prioridade ao digital, um novo ciclo de inovação para jornais impressos. Insular, Florianópolis (2018)

Lucena Filho, G.J.: Competências conversacionais: um diferencial no gerenciamento de projetos. MundoPM **6**, 75–80. ISSN: 1807-8095 (2010)

Lucena Filho, G.J.: Sobre conversações na Fundação Oswaldo Cruz (Fiocruz). Programa de desenvolvimento organizacional da Fiocruz. Brasília. 1–6 (2014)

Lucena Filho, G.J., Carneiro, A.: Competências conversacionais: uma jornada de aprendizado transformacional pelo mundo das redações jornalísticas, com base na metodologia da "Teoria Fundamentada dos Dados". In: Simeão, E., Cuevas-Cerveró, A., Botelho, R., Gómez-Hernandez, J.A. (org). Competencias em Information y Políticas para Educatión Superior: *Estudios Hispano-Brassileños*, vol. 2, 1st edn, pp. 26–38. Universidad Complutense de Madrid, Madrid. 155 p (2019)

Lucena Filho, G.J., Campos, R., Oliveira, S., Morales, M.: A technontological framework to conversations for KM: conception and potential applications. In: Proceedings of 9th International Conference on Intellectual Capital, Knowledge Management & Organisational Learning—ICICKM, Bogotá, Colômbia (2012)

Maturana, H., Varela, F.: A árvore do conhecimento: as bases biológicas do entendimento humano. Palas Athena, São Paulo. ISBN: 9788572420327 (2001)

Oliveira, E.: Relatório de Pós-Doutorado em Ciência da Computação (Completion of Post-Doctoral Course Work). Universidade de Brasília (UnB), Brasília, DF, Brazil (2017)

Santos, I.M.: Competências relacionadas à aprendizagem organizacional em equipes de projetos (Masters thesis). Universidade Católica de Brasília (UCB), Brasília, DF, Brazil. Available at: https://bdtd.ucb.br:8443/jspui/handle/123456789/1563 (2008)

Searle, J.R.: Actos de habla. Cátedra, Madrid. ISBN: 9788477110675 (1980)

Silva, J.A.B., Pereira, M.L., Ribeiro, R.S.: Estudo de caso das transformações no perfil do jornalista. Brazilian Journalism Research 9(2), 50 (2013)

Tárcia, L.P.T.: Ação, pesquisa e reflexão sobre a docência na formação do jornalista em tempos de convergência das mídias digitais (Masters thesis). PUC-MG, Brazil (2007)

Winograd, T., Flores, F.: Understanding Computers and Cognition: A New Foundation for Design. Ablex, Norwood, NJ. ISBN: 978-0201112979 (1998)

Ana Cristina Carneiro dos Santos Ph.D. in Information Science at the University of Brasília with sandwich period at Brunel University London. Master in Knowledge Management and Information Technology at the Catholic University of Brasília, with specialization in Education at the Federal University of Rio de Janeiro, extension in Project Design and Analysis by the Getulio Vargas Foundation, specialization in Planning and Systems Analysis by the Educational Union of Brasília and graduated in Data Processing Technology from the Paulista University. Acts as an undergraduate and graduate teacher. She is a specialist in Complexity Management with an emphasis on Language Ontology, as well as PMP (Project Management Professional) certified and consultant in organizational management. Her main areas of interest and current work focus on teaching, research and consultancy activities related to the processes of articulating actions and obtaining results in the networks of organizational commitments.

Gentil José de Lucena Filho University professor for 38 years, is a Bachelor of Physics (1972) and an M.Sc in Systems and Computing (1974) from UFPB (1974) and a Ph.D. in Systems Design (1978) from the University of Waterloo / Canada (1978). He did a post-doctorate in Artificial Intelligence, at the New University of Lisbon, Portugal, and has specialization courses in Strategic Planning and in Group Dynamics (by SBDG). From 1973 to 1998, he was Adjunct Professor in the area of Computer Science at the Federal University of Campina Grande (UFCG) and at the University of Brasília (UnB), a period in which (from 1981 to 1998) he was also a Senior Science and Technology Analyst at CNPq. From 1998 to 2011 he was Adjunct Professor and Researcher at the Catholic University of Brasília (UCB), in the Master in Knowledge Management and Information Technology. He has been a certified ontological coach since 1997, by the former The Newfield Group, in partnership with the Technological Institute of Higher Education of Monterrey (ITESM). He is currently Chief Researcher at the Conversational Intelligence Laboratory (LabCon). His areas of research today include: Organizational Learning, Conversations in Organizations, Ontological Coaching, Knowledge Management and Team Performance. Specifically, at Labcon, he is dedicated to developing and disseminating the research and the practical application of coaching and conversational skills in Brazilian institutions.

Gheorghita Ghinea is a Professor in Mulsemedia Computing in the Department of Computer Science, at Brunel University. Dr. Ghinea's research activities lie at the confluence of Computer Science, Media and Psychology. In particular, his work focuses on the area of perceptual multimedia quality and how one builds end-to-end communication systems incorporating user perceptual requirements. Dr. Ghinea has applied his expertise in areas such as eye-tracking, telemedicine, multi-modal interaction, and ubiquitous and mobile computing, leading a team of 8 researchers in these areas. He has over 300 publications in his research field and is the Editor in Chief of the International Journal of Pervasive Computing and Communications. Currently, his research pursuits are centered on extending the notion of multimedia with that of mulsemedia a term which he has put forward to denote multiple sensorial media, ie. media applications which engage three or more of the human sense. His work has been funded by both national and international funding agencies and has been covered by the BBC, Telegraph, and Forbes magazine, among others. He consults regularly for both public and private institutions in his areas of expertise.

Lillian Maria Araújo de Rezende Alvares Professor at the Faculty of Information Science at the University of Brasília (UnB), since 2006, where she held the positions of Coordinator of the Undergraduate Course in Museology, Coordinator of the Graduate Program in Information Science and Coordinator of the Undergraduate Course in Archivology. Post-doctorate from the School of Technology in Experimental Sciences at the Jaume I University, Spain. Ph.D. in Information Science from the University of Brasília and from the Southern University of Toulon-Var in a regimen of shared tutelage. Specialist in Competitive Intelligence from the Brazilian Institute of Information in Science and Technology. Master in Library Science from the University of Brasília and Graduated in Mechanical Engineering from the same university. The main areas of research are information management, knowledge management and competitive intelligence.

Ébida Rosa dos Santos PhD in Communication from the University of Brasília (UnB). Master in Journalism from the Federal University of Santa Catarina (POSJOR UFSC). Graduated in Social Communication, with a degree in Journalism, from the Federal University of Santa Maria. Professional Registration: Mtb 16154. Member of the Science Technology and Politics research group (CTPOL) and the Media and Politics Studies Center (NEMP).

Maria de Fátima Ramos Brandão Ph.D. in Social, Work and Organizational Psychology from the University of Brasília (2009), Master in Computer Science from the Federal University of Rio Grande do Sul (1984), Specialist in Organizational Coach from Newfield Consulting and UCB (2001) and graduated in Data Processing by UnB (1978). She is a founding professor of the Department of Computer Science at the University of Brasília, where she has been working since 1983. She coordinated the implementation of the Bachelor's Degree in Computer Science and the first Licensed Teacher's Degree in Computer Science in Brazil. Participated in the Evaluation of the PROINFO Program (MEC/SEED) and in the implementation of the National Higher Education Evaluation System (SINAES) at INEP. She coordinated the evaluation of the Casa Brazil Project, a federal government action for the digital and social inclusion of MCT/SECIS. Coordinates the Extension Network Program for Digital Inclusion—REID, started in 2010 at UnB. Worked in the production and offering of disciplines for UAB/UnB and as a teacher in the Professional Master's program in Sustainable Management of Lands and Indigenous Peoples at the Center for Sustainable Development and in the Professional Master's in Information Security at the Department of Computer Science at UnB. She served as Technical Director of Graduation of the Dean of Undergraduate Education and Institutional Prosecutor of UnB from 2013 to 2016. She is the Technical Coordinator of the MDM Research Project - Multimodal Digital Media under the International Cooperation Program CAPES—PVE Subprogram 2014.

Big Data and the Studies of Communication in Brazil

Notes on Reconfiguring a Knowledge Field

Márcio Carneiro dos Santos

Abstract The impact of the transformations generated by the digital media ecosystem in the field of Communication is discussed, based on the quantitative explosion of emitters, which are sustained by the ubiquity of networks and technological devices to support content production. Using proposals from Design Science and from the Digital Methods approach, the possibilities of epistemological and methodological expansion are analyzed, based on interdisciplinary initiatives and on the incorporation of new skills in the training of professionals and researchers, in order to face the current situation of excess data and of poorly adapted tooling available to understand it.

Keywords Data · Algorithms · Communication · Media organizations

1 Introduction

The field of Communication developed from the epistemological choices of the Humanities and Social Sciences and has incorporated approaches, methods and a mode of operation in which description and, especially subjective interpretation, have always been in first place.

Even today, the functions of prediction and prescription are strange to the descendants of this scientific tradition. Those functions are usually more oriented forms of doing projects, facing real problems and proposing solutions on other areas, such as Design, Engineering, Medicine and Science Earth.

The changes from the spread of digital processes on the contemporary informational ecosystem brought, however, a whole new set of study objects and important issues for Science Communication.

The scenario where such phenomena occur is characterized by processes that present speed, variety and volume of the information produced as main features

M. C. dos Santos (✉)
Federal University of Maranhão, São Luís, Maranhão, Brazil
e-mail: marcio.carneiro@ufma.br

© The Author(s), under exclusive license to Springer Nature Switzerland AG 2022
B. Medeiros Neto et al. (eds.), *Digital Convergence in Contemporary Newsrooms*,
Studies in Systems, Decision and Control 370,
https://doi.org/10.1007/978-3-030-74428-1_7

(González-Bailón 2013; Lewis and Westlund 2015; Lima Jr. 2012; Marth and Scharkow 2013). This fact makes certain approaches unfeasible and, as a result, they do not contribute much to the understanding simply because they cannot identify, register, nor even look for similarities or divergences for the classification.

Small or hand-treated samples can do little on movements of thousands or millions of human actions, often carried out almost synchronously, such as a group of people posting tweets on a topic that momentarily attracts collective attention (Santos 2013). No matter how dedicated is the researcher nor the amount of hours he dedicates to the collection of this data, treating them without computational help will offer little potential to draw inferences or even to capture what is actually happening.

The investigative journalists and professionals in the organizational environment were the first ones who had to deal with such problems. The data from transparency portals and hidden stories behind the numbers; the metrics of the indicators related to the presence on social media platforms; the information revealed by the monitoring tools; the flood of data from analytics solutions; all this brought to the field not only a new set of problems, but also a chain reaction that began to impact the need for new skills from professionals, different forms of approach, the search for new business models and, why not, the theoretical and epistemological review and readjustment of knowledge as well, which had been built in an digital world, quite different from the current one.

Strange as it may seem, it was the complexification of human behavior that made necessary to insert machinic processes in areas of knowledge where such approaches have never been common or welcome before. The Digital Humanities (Moretti 2005; Lemos 2002), Computational Linguistics (Santos 2014), Automated or Data-Driven Journalism (Bradshaw 2014; Bruns 2013; Rodrigues 2009), Mathematics Applied to Sociology (Bonacich and Lu 2012) and so many new hybrid forms of knowledge challenge those who propose to study contemporary processes, which are, today, supported by networks and binary machines, generating digital objects with their own characteristics and ontology (Manovich 2001), unable to be unveiled without a reorientation of research methods, tools and techniques (Moretti 2005; Van Dick 2013; Santos 2015).

Public empowerment sustained by digital media has created an explosion of broadcasters reconfiguring the networks of dissemination of information from the digital world, which were previously concentrated in large hubs of attention, such as major media outlets and official sources.

Similar to a city where several new avenues were opened, the circulation of data gained speed and volume that were never seen before, engendering processes which could only be understood through theories of Complexity, Networks and Games, all very strange to the traditional Communication researchers.

2 Reconfiguration Topics

Considering the wide range of problems to be solved and the limitations of this text, we have chosen to organize the required scientific action based on the following broad lines of action related to this topic:

(a) Epistemic expansion—incorporation of activities or attempts of prediction and prescription on research initiatives linked to the classes of problem of the real-world; encouragement to applied research and guidance for data-driven approaches, which are present in proposals as the Design Science;
(b) Methodological Extension—experimentation and testing of approaches such as the ones from Digital Methods, based on the premise of a specific ontology of binary entities that impacts their forms of apprehension, as, for example, in the ARS methodology of social network analysis;
(c) Interdisciplinarity—openness to the connection and operation of groups of researchers from different academic traditions to face problems that also arise from the hybridization of themes and in-depth complexity;
(d) Theoretical review—adaptation and proposition of constructs, models and explanations with the possibility of operation in the current contemporary information ecosystem;
(e) Diversification of skills—opening to the approach of more consistent quantitative approaches that include data analysis and data visualization, statistics and even, in some cases, mathematics and programming languages such as Python;
(f) Update on training programs for professionals and researchers—which implies on review (or renewal of proposals) in subjects, menus, programs and, eventually, areas of concentration and lines of research, in the case of graduate programs.

It must be emphasized that, in no way, the insertion of these changes should happen on an uncritical or contemplative manner, neither from a simplistic view that the approximation with data, algorithms and computational tools alone will save us all. Today, in Brazil, there is already (albeit embryonic), a group of researchers who have explored, for example, the bias and errors that the use of artificial intelligence tools (specifically machine learning) have generated in processes which ended up implying on invasion of privacy, distortion of facts and even on racism and xenophobia, as shown by Amadeu (2019) and Silva (2019).

Nevertheless, it is precisely so that for the critical tradition of the field of Communication continues to be applied on the digital environment, it is necessary to deal with these themes directly or, at least, indirectly, through work with interdisciplinary teams, which will be able to work more effectively on such matters. Indeed, being apart of these themes would place us on positions of scientific fragility and even irrelevance, on a distant movement from the essential processes guiding the transformations in the digital environment.

It is also important to remember that the identification of the transformations caused by structural changes on information systems of digital/binary essence has been made by several theorists from Communication and other areas, such as in

Castells (1999), Chwe (2000), Feenberg (2002), Lemos (2002), Santaella (2003), Vilches (2003) among others, for at least 20 years. This fact makes us believe that, after all this time, confirming or repetitive studies of these propositions have little to add to the expansion of their respective scientific fields. If such phenomena exist and have already been pointed out, it is important now to study how they occur, what patterns they follow, what forces move them and what impacts they may generate in the near future. The profusion of case studies, so common in our area, indicates a low potential for inferences and propositions of wider scope, mainly in the development of new theories and revision of the previous ones.

3   Design Science and Digital Methods

The items a and b from the list above can be addressed, among other possibilities, through the approximation with two types of approaches poorly known in the national environment of research in Communication. We have been working at LABCOM—Media Convergence Laboratory[1] with the dissemination of these theoretical and methodological possibilities, not only in terms of the development of projects oriented to applied research, but also using documented reflections, which we will be summarized below.

3.1   Design Science (DS)

The term "science of design", which was later called "design science", was introduced by the economist and philosopher Herbert Simon (1996) in a work considered seminal for this field, "The sciences of the artificial", first published in 1969. In the book, the author begins to sketch a new epistemological paradigm that today is characterized by the orientation to the solution of problems, either through the creation of new artifacts (a concept that we will detail later), or through the improvement of existing solutions. Initially focused on the fields of Engineering and Information Systems, design science has seen its use expand to Management, Education, as well as Applied Social Sciences in general, offering an alternative path for researchers who wish to go beyond the phases of description and analysis of research topics previously given.

The prescriptive and purposeful character of this design science seeks to integrate projects that, while maintaining the rigor of traditional scientific methods, also

---

[1]Founded in 2010, LabCom is connected to the Department of Social Communication at the Federal University of Maranhão (UFMA). It is dedicated to developing projects around the connections between Communication and Technology, emphasising applied research and experimentation. More information at: www.labcomdata.com.br; https://www.facebook.com/labcomUFMA/; and https://www.labcomdata.com.br/.

seek the social relevance of their findings in the implementation of objective improvements to problems of a certain class.

Design science (DS) proposes a kind of extension, on the direction of solving real problems and improving existing artifacts. In Gibbons (1994), Le Moigne (1994), March and Smith (1995), Romme (2003), Walls et al. (1992) we find several references to this view. In Brazil, the work of Dresch et al. (2015) is a reference on the topic.

The direction of the DS towards objective criteria in problem solving can be understood based on its fundamental concepts, that of pragmatic validity that "seeks to ensure the usefulness of the proposed solution to the problem. It considers: cost/ benefit of the solution, particularities of the environment in which it will be applied and the real needs of those interested on the solution"[2] (Dresch et al. 2015, p. 59).

3.1.1 Artifacts and Classes of Problems

Another fundamental concept for DS is the **artifact**. Design Science is the "science that seeks to consolidate knowledge about the design and development of solutions to improve existing systems, solve problems and create new artifacts" (Dresch et al. 2015, p. 59).[3] The concept of artifact can be understood as the final product of the route proposed by DS and, therefore, something that is associated with the specific context of the problem to be solved. The artifact, created by man, represents an intermediary between a set of knowledge established in a given area and the specific conditions surrounding the problem that the artifact must solve.

Artifacts can be divided into categories and one of the most accepted classifications is that of March and Smith (1995), who propose four types: constructs, models, methods and instantiations.

- *Constructs*—the most basic elements on the development of DS, they are conceptual elements whose objective is to establish a set of definitions used in the solution of the problem, constituting a kind of vocabulary about a certain field where such problem is inserted. These are the concepts on which the solution operates and that the researcher will use to evolve from the purely abstract to the tangible (Fig. 1) and applied to the given situation.
- *Models*—descriptions of a certain system that establish relationships between constructs. They are a kind of representation of reality that seeks to describe it, even though through simplifications, but it aims to understand its internal operating logic for using it as a solution.
- *Methods*—sets of procedures and actions oriented towards the performance of a given task or solution to a given problem. The methods can be related to established models, being one more step in the scale between abstraction and tangibility of the solution that we have proposed previously.

---

[2]Author translation.

[3]Author translation.

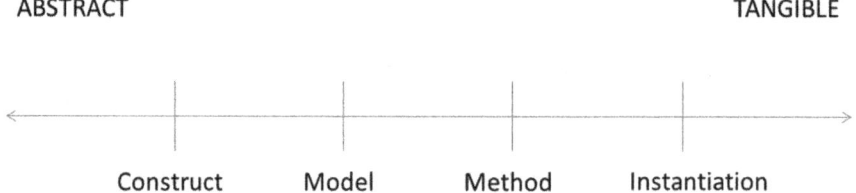

Fig. 1 Scale of tangibility of artifacts in the DS. *Source* Santos (2015)

- *Instantiations*—the concept of instantiation, well known among programmers and computer scientists, in DS represents perhaps the most tangible level of the solution created on the previous context that inspired it. In other words, it represents the artifact in operation in the environment that generated the need of the solution. The instantiations also allow us to evaluate something important within the DS proposal, which is its effectiveness in relation to the proposed problem or the intended improvements in the existing system (Fig. 1).

A fifth type of artifact is admitted by some authors who refer to it using terms such as "technological rules", "design rules" or more commonly "design propositions".

- *Design Propositions*—These design or project propositions would be theoretical contributions that can be made from the application of the DS principles on face of a specific type of problems, or in their own terms, operating on a class of problems.

The term "class of problems" that we have used is also part of the important concepts of DS. Sets of practical or theoretical problems that have established a set of solutions or artifacts linked to them constitute a class of problems. As an example of Communication and Social Sciences, we could mention the general need of collecting data in repositories on the internet, which we could name as digital data collection. Whether for the production of a journalistic story, for a management plan or for the definition of a public policy on a given topic; with the digitization processes and the growing use of databases, the need of obtaining such information, accessing their repositories available on the network, such as transparency portals, for example, characterize a class of problems on which artifacts operate, such as scraping[4] and automated extraction methods, as well as the

---

[4]Santos (2015) http://www.filosofiacienciaarte.org/attachments/article/987/MCSLABCOM-RevObservatorioPaper.pdf. Digital methods and the memory accessed by APIs: Development tool for extracting data from journalistic portals with the WayBack Machine. We explore the possibility of automation of data collection from web pages, using the application of customized code built in Python programming language, with specific HTML syntax (Hypertext Markup Language) to locate and extract elements of interest as links, text and images. The automated data collection, also known **as scraping** is an increasingly common feature in journalism. From the access to the digital repository site www.web.archive.org, also known as WayBackMachine, we develop a proof of concept of an algorithm able to recover, list and offer basic tools analysis of data collected from the various versions of newspaper portals in time series.

available instantiations exemplified by the algorithms in a given programming language, which operate to solve such problems. In this last example, the codes could not only be classified as instantiations but also as methods, since they execute sequences of commands to perform their functions.

3.2   Digital Methods

Rogers (2013) states that even using traditional methods on digital research, we may, in some situations, be using an inappropriate tool.

> For example, data scanning and extraction, collective intelligence and classifications based on social networks, even if from different genera and species, are all internet-based techniques for collecting and organizing data. Page Rank and similar algorithms are means of ordering and ranking. Word clouds and other common forms of visualization make relevance and resonance explicit. How could we learn from them and from other methods online to reapply them? The purpose would not really be to contribute to the refinement and construction of a better search engine, a task that should be left to Computer Science and related fields. Instead, the purpose would be to use them and understand how they handle hyperlinks, hits, likes, tags, datestamps and other natively digital objects. Thinking about these mechanisms and the objects with which they can deal, digital methods, as a research practice, contribute to the development of a methodology for the environment itself (Rogers 2013, E-book).

The proposal of Rogers is in line with the path we now propose starting from a view of the contemporary world where digital has a growing centrality, composed of entities with specific characteristics and, therefore, also requiring an adaptation or methodological extension capable of collaborating with research whose objects somehow have this trait.

Thus, we define digital methods as the set of tools, processes and research approaches that consider the ontology of digital objects and the network structures through which they circulate, using intensive computational resources for data collection and analysis.

Such solutions offer a kind of scale of use, as represented in the graph (Fig. 2).

Such scale ranges, at an initial level, from the use of pre-existent tools and techniques included on the default setting; at a medium level, to adjustments for customizing them to meet our specific needs; or, at a higher level, through the creation of solutions based on programming and code development.

In the pyramids above we exemplify the scale in a situation of data collection using Google search, initially with its normal interface, then from a solution with greater personalization power, such as alerts,[5] finally through a specific code to collect and store that data.

---

[5]https://www.google.com/alerts.

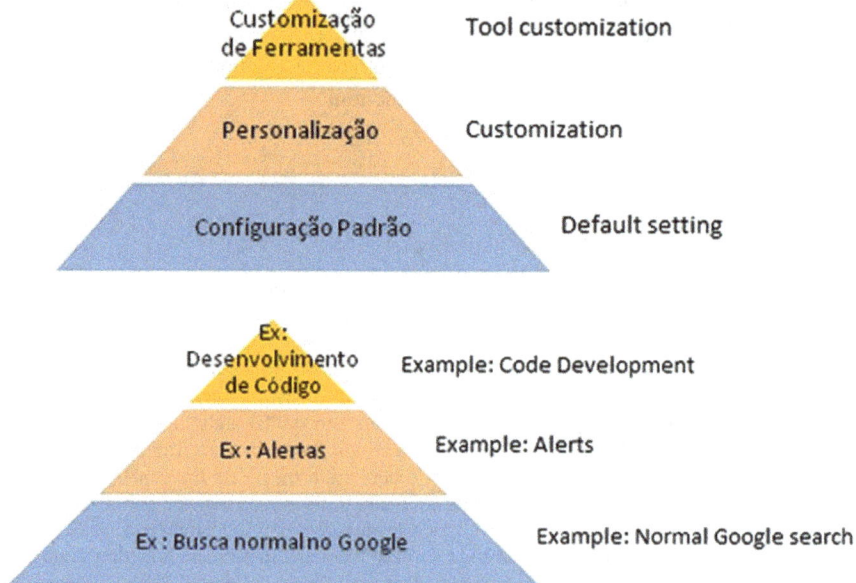

Fig. 2 Representation of the scale of use of digital methods. *Source* Santos (2015)

In general terms, the approach we propose boils down to the following steps:

1. Identify the structure containing the data we need. Some possibilities often appear:

   (a) Databases that allow friendly queries by filling out simple forms or procedures. Example: government transparency portals where it is possible to request data on a specific topic and period.

   (b) APIs that require structured requests on the format they establish, that is, respecting their own syntax. Example: Twitter and Facebook APIs, which need a specific application to request content, such as the applications that access them on our cell phones, or a custom code that can establish such a dialogue and collect the information delivered by API for each type of request.

   (c) Content available on the web that can be extracted directly via scraping techniques. For instance, texts of articles in news portals or tables and general information published, such as weather forecast, dollar quotation and sports competition results.

   (d) Information protected in closed environments, accessed only by registered users and that have protection mechanisms, such as data encryption and others. Such environments can eventually be accessed by hacking techniques, which are out of the scope of this text.

2. Format the query or data request in line with the type of repository where they are located according to the options above.
3. Analyze the data collected from the possible processing based on what was actually achieved.

We can then combine on a table the scale of use from Fig. 2 with the various forms of the most common data structuring. Crossing four of the most common forms of repositories structured in online databases, we list some possibilities of digital methods application in its three levels (Table 1).

Table 1 Matrix of collection possibilities by digital methods depending on the form and location of the data and the levels of application

Data structure	Initial level: standard tools	Medium level: custom tools	High level: code development
(a) Open database and repositories	Requesting data through the database interface, receiving the result on a standard delivery format. Ex: Access to IBGE's SIDRA database and download the file in Excel format or in CSV	Use of filters and resources of analysis and visualization offered by the platform, changing the way the result is delivered according to the options offered. Ex: Use of SIDRA's advanced functions and graph generation	Code to automate access to the database by making successive and customized requests, collecting and saving the data in another type of data structure or format. Ex: Python with Splinter or Selenium modules
(b) Servers with access through a specific API	Access through the platform's official application or through its standard web page. Ex: Using the FB app on your cell phone or accessing the page: www.facebook.com	Access through third party applications that also access the platform server, but offer additional functionalities. Ex: Node XL	Code to access the API platform directly, by collecting all the information provided by it and also making successive requests capable of collecting larger volumes of data
(c) Content on HTML web pages	Google search, manual access and eventual collection through CTRL + C and CTRL + V	Use of specific scrapping tools Ex: Portia, Web Scraper	Code development for collection and analysis. Ex: Python with the Beautiful Soup module
(d) Protected data through login access	Access through registration request and normal log in	Tools of general hacking, such as brute force or social engineering	Development of invasion codes as worm or trojan

4 Final Consideration

The need for a reconfiguration of research practices in the field of Communication seems to be an important topic to us when dealing with objects and research problems on the context of the contemporary information ecosystem, constituted from binary entities, hyperconnected in increasingly complex networks.

The characteristics of volume, variety and speed directly impact the effectiveness of the approaches that are not oriented to deal with their presence, which interfere in the results, reduce the possibilities of more solid inferences and, even, prevent the apprehension of increasingly important phenomena in the list of Communication researchers.

The proposals of Design Science and Digital Methods, cited here only as examples, constitute a possible (but not unique) path capable of adding new horizons in epistemic and methodological terms to researchers in a field increasingly challenged by the overwhelming production of data and content, generated not only by official sources and major communication vehicles, but also by an explosion of emitters empowered by mobile devices and expanding internet infrastructure in the country.

Obviously, It is not suggested here the abandonment of a whole set of consolidated knowledge that made the national field of Communication set up and grow over decades. On the contrary, we understand that an expansion reflecting the current centrality of networks and digital entities is necessary based on what has already been done. However, this has to be done without blocking the new possibilities that interdisciplinarity and teams of researchers from different academic backgrounds have to offer in many research initiatives facing the scenarios described in this text.

It is not an easy task, but through the lines here pointed out: epistemic expansion; methodological extension; interdisciplinarity; theoretical review; diversification of skills and updating of training programs for professionals and researchers; we understand that it is possible to take the first step.

Such paths are not presented here as a single and infallible solution, but as metaphors or guides for an internal movement of reconfiguration and review of research practices, which are urgent, necessary and, in our view, inevitable.

References

Amadeu, S.: Democracia e os códigos invisíveis: como os algoritmos estão modulando comportamentos e escolhas políticas. Coleção Democracia Digital. Edições Sesc, São Paulo (2019)

Bonacich, P., Lu, P.: Introduction to Mathematical Sociology. Princeton University Press, Princeton (2012)

Bradshaw, P.: Scraping for Journalists. Leanpub (2014) [E-book]

Bruns, A.: Faster than the speed of print: reconciling 'big data' social media analysis and academic scholarship. First Monday **18**(10), 1–5. Available in http://firstmonday.org/article/view/4879/3756 (2013). Accessed at 23 July 2017

Castells, M.: A sociedade em rede (Tradução: Roneide Venâncio Majer). Paz e Terra, São Paulo (1999)

Chwe, M.S.Y.: Communication and coordination in social networks. Rev. Econ. Stud. **67**(1), 1–16 (2000)

Dresch, A., Lacerda, D.P., Júnior, J.A.V.A. Design science research: método de pesquisa para avanço da ciência e tecnologia. Bookman Editora (2015)

Feenberg, A.: Transforming Technology: A Critical Theory Revisited. Oxford University Press, New York (2002) [E-book]

Gibbons, M. (ed.): The New Production of Knowledge: The Dynamics of Science and Research in Contemporary Societies. Sage (1994)

González-Bailón, S.: Social science in the era of big data. Policy Int. **5**(2), 147–160 (2013). Retrieved at: https://pdfs.semanticscholar.org/6e78/b1133713cb17aabbc3bf421a6e51bc538eca.pdf. Acessado em 23 July 2017

Le Moigne, J.: Le Constructivisme - fondements. ESF, Paris (1994)

Lemos, A.: Cibercultura, p. 320. Sulina, Porto Alegre (2002)

Lewis, S., Westlund, O.: Big data and journalism—epistemology, expertise, economics and ethics. In: Digital Journalism, vol. 3, pp. 447–466 (2015)

Lima Jr., W.: Big data, jornalismo computacional e data jornalismo: estrutura, pensamento e prática profissional na Web de dados. In: Estudos em Comunicação nº 12, págs. 207 a 222. Covilhã: UBI, 2012. Retrieved at: http://www.ec.ubi.pt/ec/12/pdf/EC12-2012Dez-11.pdf (2012). Accessed at 07/21/2017

Mahrt, M., Scharkow, M.: The value of big data in digital media research. J. Broadcast. Eletronic Media **57** (2013). Retrieved at: http://www.tandfonline.com/doi/abs/10.1080/08838151.2012.761700

March, S.T., Smith, G.F.: Design and natural science research on information technology. Decis. Support Syst. **15**(4), 251–266 (1995)

Manovich, L.: The Language of New Media. MIT press (2001)

Moretti, F.: Graphs, maps, trees: abstract models for a literary history. Verso, Brooklyn (2005)

Rodrigues, A.A.: Infografia Interativa em Base de Dados no Jornalismo Digital. 130f. Dissertação (Mestrado em Comunicação)—Universidade Federal da Bahia, Salvador (2009)

Rogers, R.: Digital Methods. MIT Press, Cambridge (2013) [E-book]. https://drive.google.com/file/d/0BwblN2uXiXNjQnNMOFFUQjc2enM/view. Accessed at 21 July 2017

Romme, A.: Making a difference: organization as design. Organ. Sci. **14**(5), 558–573 (2003)

Santaella, L.: Da cultura das mídias à cibercultura: o advento do pós-humano. Revista Famecos **10**(22), 23–32 (2003)

Santos, M.: Conversando com uma API: um estudo exploratório sobre TV social a partir da relação entre o twitter e a programação da televisão. Revista Geminis, **4**(1), 89–107 (2013). Available in: www.revistageminis.ufscar.br/index.php/geminis/article/view/129/101. Acesso em: 20 abr. 2013

Santos, M.: Textos gerados por software. Surge um novo gênero jornalístico. In: CONGRESSO BRASILEIRO DE CIÊNCIAS DA COMUNICAÇÃO, 37. 2014. Anais..., Foz do Iguaçu. Available in: http://www.labcomufma.com/biblioteca-digital (2014). Accessed at 26 Jan 2014

Santos, M.: Métodos Digitais: a internet e as redes como instrumentos de pesquisa. In: CONGRESSO BRASILEIRO DE CIÊNCIAS DA COMUNICAÇÃO, Anais V Colóquio Brasil-Argentina, Rio de Janeiro. Available in: http://portalintercom.org.br/anais/nacional2015/lista_area_COL.htm. (2015). Accessed at: 07/29/2016

Silva, T.: Linha do Tempo do Racismo Algorítmico. Blog do Tarcízio Silva. Available in: https://tarciziosilva.com.br/blog/posts/racismo-algoritmico-linha-do-tempo (2019). Accessed at: 20 Aug 2019

Simon, H.: The Sciences of the Artificial, 3rd edn. Mit Press, Cambridge (1996)

Van Dick, J.: The Culture of Connectivity: A Critical History of Social Media [E-book]. Oxford Press, New York (2013)
Vilches, L.: A migração Digital. Loyola, São Paulo (2003)
Walls, J.G., Widmeyer, G.R., El Sawy, O.A.: Building an information system design theory for vigilant EIS. Inf. Syst. Res. **3**(1), 36–59 (1992)

Márcio Carneiro dos Santos Doctor by the Digital Technologies and Digital Design (TIDD) program at PUC-SP. Postdoctoral internship at UNB in the line of research Theories and Technologies of Communication. Permanent professor of the Graduate Program in Design at UFMA. Permanent Professor of the Postgraduate Program in Communication—Professional Master's Degree at UFMA. Adjunct Professor in the Department of Social Communication in the area of Journalism in Digital Networks. He is coordinator of LABCOM—Media Convergence Laboratory. Leader of the CNPq research group—Technology and Digital Narratives—TECND. Deputy coordinator of the Digital Content and Technological Convergences GP at INTERCOM Nacional. Coordinator of the Working Group on Digital Communication and Pen-sacom Technologies—Brazilian Communication Thought. Has published works in the areas of Intelligent Systems Applied to Journalism, Narratives in Digital Environments, Immersive Journalism, De-sign Science, Network Theory, Analysis of Social Networks and Philosophy of Technology. Adelmo Genro Filho Award, from the Brazilian Society of Journalism Researchers—SBPJor, in 2018, in the Applied Research category. FAPEMA 2011 Award in the Technological Innovation category. Participates in the research group ComTec—Communication and Technology and the JorTec Network—Journalism and Technology. Master in Communication from Universidade Anhembi Morumbi—São Paulo. Marketing Specialist from ISAN/FGV-Rio. Graduated in Social Communication from the Federal University of Maranhão in the qualification of Journalism. He worked for 20 years in the area of audiovisual content production as an image director, editor, screenwriter and executive producer.

Disinformation Culture

A Reflection on Origins, Dissemination Methods, Contexts, and Confrontation

Gislane Pereira Santana and Elmira Luzia Melo Soares Simeão

Abstract Situations that promoted the distortion of facts at different moments in history demonstrate important aspects and arguments for a discussion on the theme of "disinformation", in the context of Information Science. A debate is proposed on initiatives and measures that can help raise people's awareness, improving their training for choosing and using information in different situations, notably on social networks. This chapter aims to discuss possibilities and studies that can help to halt the spread of false news on social media and a culture of disinformation. As a way of coping with the problem, the authors suggest the adoption, as a reference, of the concepts of Digital Literacy and Information Skills (CoInfo).

Keywords False news · Disinformation culture · Disinformation · Digital literacy · Competence in information

1 Introduction

In the debate about the current problem of production and proliferation of false news on social networks, it is necessary to observe the phenomenon as a reproduction of old practices, and also treat it as a process of "social disinformation". The term "disinformation", in the context of Information Science is widely discussed, and its characteristics are the subject of studies in the area of Communication and Computing and can be added to this debate. It is based on the assumption that Information Science can indicate paths, and contribute to contain the advance of "fake news" (an expression considered controversial by Unesco) and that it may stimulate other areas for investment in the practices of Information

G. P. Santana (✉) · E. L. M. S. Simeão
Universidade de Brasília, Campus Darcy Ribeiro, Brasilia 70910-000, Brazil
e-mail: santana1204@gmail.com

E. L. M. S. Simeão
e-mail: elmira@unb.br

Competences (CoInfo), linking ethical concepts in this Digital Literacy and proposing the ethical use of technologies and training in schools and universities.

Ancient records show that mankind has always used false arguments to make real what, in fact, does not exist. Or try to camouflage real facts for the benefit of interests that are not always declared and publicly acceptable. Currently, with the popularization of the term "Fake News", and the use of the expression FAKE to indicate the falsity of any fact or situation, it is essential to investigate how information, in its multiplicity of possibilities, could be falsified, showing a "Culture of Disinformation" in the social practices revealed through the use of technology.

The professor and director of the School of Communication and Arts at the Federal University of Rio de Janeiro (ECO-UFRJ), Ivana Bentes, claims that there has always been false news, rumors and facts in the media used with the intention of destroying reputations, however, because internet and social networks, those false news gained speed of diffusion and became automated and massified (Bentes 2018). Bentes brings this reality to the debate as a "culture of disinformation".[1] The more misinformation incorporated in the messages, the more disorderly consumption, often amplified by robots, causing individuals to disseminate "fakes" without realizing the negative impact on the communication of social networks and media. This dynamic is part of the culture of disinformation and it is advisable to allow the citizens a more critical and prepared attitude for the use of alternative sources, especially when one is not sure of the origin of the information, or when reading or research is not considered aspects of reliability of the information source.

In the theoretical framework on studies in the area of information science and disinformation, analysis is observed from two perspectives, one technological and the other educational. In the use of information and communication technology. It is possible to fight false news deploying technical instruments and software. On the other hand, and perhaps the most challenging aspect, is to correct attitudes of social network users developing the necessary skills to interpret the quality and truthfulness of the facts that are made available and disseminated in news format. To face the misinformation, the use of technological resources demands the creation of new forms of learning and digital literacy is presented as the main premise of the moment.

2   History of Disinformation

The internet is a fertile field for the appearance of False News and a space for crimes such as defamation and others (Darnton 2017). In the 2018 report, "A short guide to the history of fake news and disinformation", prepared by the International Center for

---

[1]News provided by Ivana Bentes, professor and director of the School of Communication at the Federal University of Rio de Janeiro (ECO-UFRJ). In the Journalism seminar: the new configurations of the Fourth Power. São Paulo, SESC Vila Mariana, August 2018.

Journalists ICFJ, aspects are described presenting the problem in the context of journalistic production and the mechanisms of protection and care with information (Posetti and Matthews 2018). This document points out a summary of what is meant by information disorder, reporting interesting historical facts that served as a basis, evidencing a "Culture of Disinformation". The report is focused on the evolution of events over time, beginning its analysis in the fourth century B.C.

It can be seen in the ICFJ report (Posetti and Matthews 2018) and in (Darnton 2017) that already in the dawn of civilization, information as an instrument for "defamation" gained prominence. In ancient Rome, false reports were used as a political strategy to dishonor Emperor Mark Antony, who was accused of disrespecting the empire by becoming the mistress of the Egyptian queen Cleopatra. Enemy senators used pejorative terms against the emperor to cause instability and distrust in the empire. In the political dispute of that time, the means used to defame opponents was the recording, in coins, of phrases that tarnished the reputation of the opponents. It can be said that the technique is still widely used today (along with other instruments) and has gained strength with the possibility of the proliferation of false news on social networks and in political disputes and elections. In Rome, in 59 BC. the spread of news in Acta Diurna, an efficient communication channel at that time, was based on a desire by Emperor Julius Caesar to keep citizens informed (or uninformed?). The newspaper was written on large white signs placed in public places throughout the city of Rome carrying information that pleased the emperor. This technique of distorting the reality of the past is comparable to the deceptive political marketing used in current majoritarian campaigns that wish to perpetuate parties or candidates for public office.

In the Byzantine empire, sixth century, the historian Procopio was engaged in elaborating false information in order to denigrate the Emperor Constantine's reputation. The compilation of texts he wrote became a book of chronicles with the title "Anecdota" that could only be published after his death.

Manipulation is the term highlighted in the ICFJ (Posetti and Matthews 2018) e (Darnton 2017) report, to refer to false news in the sixteenth century. At that time, the poet Pietro Aretino, in the midst of the Italian Renaissance, manipulated the papal conclave of 1522, and entered history by showing his sonnets around a statue, Pasquino, near Piazza Navona, in Rome. Hence, the origin of the word "Pasquim" to characterize a channel that produces sensational information. Aretino wrote sonnets with false information about all candidates for the position of Pope, with the exception of Giulio de Médici, his patron. The aim was to support Médici's campaign with false information that would help elect him. The strategy, however, did not convince.

The slackers were losing space to another type of vehicle, known as "canards", which was a type of newspaper printed in large size with engravings that called the attention of the credulous. This communication vehicle was a gazette and the content of its information, rumors and false news. He was published for several years in Paris, purposely spreading slanderous facts. In this newspaper, the illustrators' creativity was audacious in printing the image of Queen Marie Antoinette's face in publications. Because of this "dubious" editorial line, the booklet gained popularity, working to spread intentionally false political propaganda. According to Darnton (2017), although it was not possible to measure the impact that the false news had on the queen's life, they certainly contributed to increase the hatred of the population culminating in the execution of Marie Antoinette in 1793.

Darton points out that in the eighteenth century, another channel use for spreading fallacious news emerged, with a new configuration that aimed to spread malicious rumors in the form of songs and poems in small passages, very similar to what we now know as tweets. These literary tweets overthrew the Count of Maurepas from the ministry. He was King Louis XVI's secretary of state and this fact changed the French political landscape of 1749. The episode was even considered as one of the causes of the French Revolution of 1789.

In the nineteenth century, the publication of a satire was used by the newspaper to make misleading claims about alien life using real information from a scientist with a piece of science fiction. This was the channel that journalists found for the dissemination of news in the newspaper "The Sun", in six articles that reported the existence of life on the moon, with the aim of deconstructing the image of the astronomer and mathematician John Herschel, recounting in a fallacious to his discoveries in 1835 Andrews (2018).

In the twentieth century, with Adolf Hitler in power and the rise of Nazism in Germany, the highlight was the creation of the ministry of propaganda and enlightenment, by Joseph Goebbels, with the aim of spreading false messages to favor the Nazi regime. The aim was to stimulate hatred against Jews, using theater, cinema and the press as a means of dissemination. The false news was produced so effectively (and together) that it distorted the facts and the atrocities committed appeared to have popular support. Consequently, they wanted to deny the existence of the Holocaust. This campaign is highlighted as the most tragic in history (Herzstein 1978), a demonstration of the strength of false propaganda and the culture of hatred as a weapon of manipulation to hold political and economic power with dire consequences.

Twentieth-century Russia was the forerunner in the creation of an organization called "troll factory" to publish a large number of messages or post on social networks with the aim of undermining the credibility of voters regarding the security of the country's electoral system. The movement also encouraged the creation of opposing groups (with false profiles), aiming to cause problems and influence public opinion. Perhaps this action was the trigger for an entire process of disseminating fraudulent information in elections today.

At the beginning of the twenty-first century, another campaign to disinform public opinion was sponsored by The New York Times in the form of sensational articles. Articles, without sources or evidence, indicated the existence of a field where chemical weapons were produced in Iraq circulated in that newspaper, contributing to strengthen the decision of the Americans to attack that country in a campaign that became known as the "War on Terror". The fraudulent action was a response to the September 11, 2001 terrorist attacks. This misinformation about weapons of mass destruction has had several consequences for the United States, provoking heated and violent discussions. The newspaper released the news that the supposed war arsenal that was being produced in Iraq was a real fact, however, without qualifying sources or proving the authenticity of the information (Miller 2001, 2003). As a way of minimizing the situation, in 2004 The New York Times apologized for spreading information about the use of weapons of mass destruction in Iraq. The newspaper would continue to publish reports aimed at correcting the false record (The New York Times 2004).

Recently in Europe, other events motivated by the spread of false news have gained prominence in the UK's exit from the European Economic Bloc. By means of a plebiscite that became known as Brexit, in June 2016, a popular consultation was made with the purpose of knowing what the British opinion in relation to the UK's permanence in that economic bloc. According to McIntyre (2018), the favorable outcome to the withdrawal of the United Kingdom from the European Economic Bloc was biased because, before the popular consultation, there was a massive campaign to influence the decision of the population, with a wide spread of rumors about Brexit in Great Britain. In order for the forged facts to reach the largest number of people, buses were used to display ads with false statistics that the United Kingdom was sending 350 million euros a week to the Bloc and that, after the eventual departure, that amount would pass to be destined to the public health service. According to McIntyre (2018), the broad campaign of disinformation about Brexit certainly influenced the outcome of the referendum.

The elections in the United States of 2016 definitely marked the phenomenon of fakes on social networks, as false news had never been so widely disseminated in all media in a coordinated way as it happened in this episode. Shortly before the start of the American referendum, it was discovered by the international media that there was a lucrative false news company controlled by teenagers, users of social networks. Headquartered in the small town of Veles, the former Yugoslav Republic of Macedonia, the group received salaries to "manufacture" and spread information. Through investigations, more than 100 pro-Trump websites were identified that spread fabricated news and these domains were all registered in Veles (Subramanian 2017).

Another fact of great repercussion that involved social networks was the use of data from millions of Facebook users, in order to misinform. In 2018, the Cambridge Analytica scandal broke out. This company, focused on strategic communication in election campaigns, combined mining and data analysis to produce false information. The objective was to create political messages targeted at a specific group of voters with the intention of influencing the 2016 US presidential elections. An aggravating factor in this episode was that Cambridge Analytica's vice president left the corporation to work on Donald Trump's election campaign (Lee 2018).

Recently another form of tampering has made use of Artificial Intelligence to edit videos. This fraud technique can falsify pornographic content, for example, or produce montages of false speeches to defame people as was done, with other techniques, in ancient Rome. The combined image and audio editing is called "DeepFake", an image manipulation feature that allows forging situations to compromise the reputation of those who are the target of this type of trap. The preferred targets of the creators of "deepfake" are politicians and celebrities.

According to Metz (2019), a journalist for CNN Business, deepfake is a combination of the terms "deeplearning" and "fake"—they are persuasive-looking but fake video and audio files. The assembly and manipulation technique uses state of art and relatively accessible Artificial Intelligence (AI). This technique can become a problem, due to the large number of videos available on the internet portraying the daily lives of personalities and politicians, and the easy access to this material that serves as a source for criminals and unemployed on duty (Metz 2019).

3   Information Science and Disinformation

Capurro states that in the rise of information science, the discussion among experts generated at least 134 notions of information, which means a wide and sometimes contradictory dimension of the term (Schrader 1986). However, the author explains that no reference to the term disinformation in a negative way and its derivations were considered. Capurro debates: "What is information science for? Like information science, conceived as a hermeneutic-rhetorical discipline, it studies the contextual pragmatic dimensions in which knowledge is shared positively as information and negatively as disinformation, particularly through technical forms of communication" (Capurro 1991, p. 10).

For Capurro, this breadth of understanding of the term indicates that readers and consumers select information based on their own mental model, cultural background, social influence, historical and ideological experiences (Capurro et al. 2007). It also depends on the interpretation skills of each individual, hence the complexity and importance of the theme. Studying information sometimes means understanding disinformation as part of the issue and facing it requires a process of investigation and interpretation of information flows, using the assumptions of Information Science for this.

For Belluzzo (2005) "the misinformation in this era is perhaps the reason for the existence of many social problems, since it affects human beings in their greatest property: rationality". It is inferred that the manipulation of information is an intentional attitude to achieve the objective on purpose. Knowledge about message production determines success in manipulating content and form. Understanding the flow allows dissemination that is a fundamental part of the manipulation process.

4   Perspective on the Use of ICT to Fight Against Disinformation

Information and Communication Technologies (ICT) disseminate diverse content and also false information that is enhanced in social media. Researcher Saracevic (1996) highlights that in the interdisciplinary aspect Information Science (IS) offers eclectic arguments assisting in the interpretation and treatment of current information based on care and a more accurate observation of contexts. For Saracevic (1996), new information technologies are definitely linked to IS, collaborating to solve problems with practices related to the selection of reliable sources. IS in its essence also presents arguments that contribute to guide the understanding and ethical reflection in the use of technological supports, since the production of messages is limited to the social, political, economic and cultural spheres. Once "IS had, and [still] has, an important role to play due to its strong social and human dimension, which surpasses technology" (Saracevic 1996, p. 42). This science uses search methods for reliability of information, and accuracy and precision of data in all social contexts.

It is necessary to emphasize the relevance of IS in combating the culture of disinformation as it is certain that relevant knowledge improves the quality of communication processes, making them more selective and user-oriented, which provides significant contributions to quality in tasks and relationships. In order to face the phenomenon of "false" the use of technologies should facilitate access and use of information services, with devices that identify the possibility of fraud or dubious information, and for this the current research must be directed. Obviously, we are faced with the challenge of necessarily multidisciplinary research. In order to make software more efficient in the task of identifying possible manipulations or inconsistencies and also to encourage ethical education for users and message producers, IS must associate with computing, education and other areas of knowledge.

It is important to remember the educational task of raising awareness on the one hand, since the simple reflection on immaterial damage is sufficient to perceive the need for the ethical use of means and messages. On the other hand, develop methods for more competent use in relation to research and information selection practices whether in the workplace, at school, in universities and in communities in general. Recalling that the identification of reliable sources and people is a permanent and lifelong learning.

Currently, at the same time that false news is "manufactured", various actions and tools are being developed to combat it. Initiatives such as the creation of websites that verify the authenticity of news known as "Fact Checking" aim to minimize the dissemination of false content information and consequently the impact it can cause in people's lives (Neisser 2015). Around the world some government and business resolutions have been taken to prevent false news from being disseminated. Software based on Semantic Technologies for text analysis and information extraction in the investigative process is another possibility as well as techniques for Natural Language Processing (NLP). One of the main objectives of NLP is to obtain intelligence from unstructured content expressed in a natural language. Almost all current research involving NLP is based on machine learning, statistical data and, more recently, on Deep Learning neural networks (Sateli et al. 2017).

The language characteristics used in the processing of false news have been explored in the area of Natural Language Processing (NLP). Rafael et al. (2018) highlights that "[…] attempts to deal with false news are relatively recent both from a theoretical and practical point of view […] some previous works have shown that human beings have a poor performance in separating true and false news, […] and that the domain can affect this but others have produced promising automatic results" (Rafael et al. 2018, p. 2).

According to Pardo (2018), it is common for people to check a news item through a message exchange application to check if it is true. Pardo (2018) points out that it is possible to do this type of checking using tools. A group of researchers from the University of São Paulo (USP) and the Federal University of São Carlos (UFSCar) developed a project that resulted in the creation of a platform, under testing, that has this objective. Pardo (2018) also highlights that, "… when a person is lying, unconsciously, it affects the production of the text. The words used and the structures of the text change. In addition, the person is usually more assertive and emotional. So, one way to detect misleading texts is to measure these character-istics" (Pardo 2018, p. 4).

To compose this context, to make it possible for machines to recognize content, information researchers had to teach them to identify what is lie and what is truth. They used artificial intelligence techniques so that the computer was able to rec-ognize in the texts characteristics that it identified as being false and true and differentiate them. They produced a set of false and true news in the Portuguese language. The researchers believe it will make it easier for the machine to evaluate future texts. Thus, all information entered by humans and the standards created by them to analyze these data sets shape operating systems to perform future tasks. To avoid possible errors and give credibility to their investigations the researchers used some parameters as criteria to validate their work such as the average number of verbs, nouns, adjectives, adverbs and pronouns identified in the texts.

Pardo (2018) highlights that the use of artificial intelligence technique to carry out this research was the solution […] "We employ classic machine learning methods, which are among the most used today, and we managed to train the system with a 90% index of correctness in the classification of news" […] the professor explains that the rate of correctness is high because the system evaluates, simultaneously, several properties found in the texts.

Although technological innovation processes are, by definition, comprehensive and encompass all natural languages most of the progress in this field, so far, is related to a single language, English (Rehm 2017). The efforts of UFSCar researchers, mentioned in this work represent another step in the search for solu-tions, demonstrating the interest of Brazilian society in combating false news.

5   Perspective of Education Under the Threat Posed by Disinformation

New skills are needed to interpret the quality and veracity of the facts that are made available and disseminated in news format. According to Leu et al. (2014), literacy has evolved with new technologies, but no technological advent has affected literacy as quickly as the Internet. According to Coiro (2008), having access to the Internet is a good thing, because having news search resources available from different sources leads users of social networks to know how to think critically, create, innovate and participate ethically in digital environments, and this represents for him a social differential.

The United Nations Educational, Scientific and Cultural Organization (Unesco) in its quest to strengthen the teaching of journalism, supported the publication of a manual for education and training in journalism, entitled "Journalism, False news and Disinformation," which is part of the Global Initiative for Excellence in Education in Journalism. This contribution brings "a set of cutting-edge knowledge that aims to get involved with the teaching, practice and research of journalism according to the global panorama, including the sharing of good international practices." It represents the suggestion of an "internationally relevant curriculum, open to membership or adaptation, in response to the problem arising from the global misinformation that confronts societies in general" (Unesco 2018, p. 7).

The manual is a contribution from Unesco and sponsors to the dissemination of the work, in an attempt to improve the training of journalists, which is the main focus of Unesco's International Communication Development Program (PIDC). This publication makes it clear that the term false news should be avoided in the context of information since these are "news" that mean verifiable information and that term takes the credit for information that represents real news.

Also highlights that misinformation and incorrect information is not quality journalism as it is oriented by ethics and professional standards. The purpose of the publication comes among an important scenario as it seeks a new posture, mainly in the training of new communication professionals, combining Communication and Information Science disciplines.

The United Nations Educational, Scientific and Cultural Organization (Unesco) created in 2013 a curriculum model for teacher training for Media and Information Literacy (MIL) with a view to disseminate this format as a reference in tackling the problem of false news. It presents ways of knowledge access to strengthen freedom of expression and characterizes the skills and essential attitudes so that there is a skilled participation and appreciation of the media and information providers.

The publication unifies concepts between information and media literacy, considering freedom of expression and access to information through ICTs, treating these terms as a combined concept. Still considering the studies by Wilson (2013), it is emphasized that the methodology provides an outline of the contents and activities that can be adapted for teachers and students according to the needs of educational institutions in their respective countries.

Competence in Information (CoInfo) as an educational practice has techniques and methods to stimulate good social practices of communication and dissemination of information and as a theoretical instance produces studies that corroborate the strategic understanding of research and information dissemination practices for individuals and institutions. It is necessary to avoid that individuals position themselves favorably in the face of situations that, in theory, would have to be refuted. Disinformation generated by false finds fertile ground in people's lack of cognitive competence to evaluate false news generated by the media and/or social networks. Having Information Competence (CoInfo) has become almost a defense mechanism and a quality differential for professionals and should be encouraged in schools and universities (Belluzzo 2014).

Competencies are skills that a person must have to perform an activity well or to conduct a research and access relevant and truthful information. Rabaglio (2001) simplifies the concepts classifying them as: (1) knowledge—related to life experience, training, etc.; (2) skills—are related to the ability to perform physical or mental activity; (3) attitude—it is related to behavior in the face of day-to-day situations and activities. It can be inferred that to perform an action the individual needs to have experience, capacity and behavior to face situations that require a coping posture. Expanding the concept in the current context of information, Belluzzo (2014) and Simeão et al. (2016) have corroborated with studies and elaboration of concepts in the use of Information Skills (CoInfo), which can help to fight the phenomenon of false news.

According to Information Science, information competence is based on a set of knowledge, skills and attitudes related to the informational context. Dudziak (2008) highlights the thematic axis of informational competence for governance and citizenship which aims to promote people's participation and control over their own actions, in addition to going beyond the search for information, but knowing why, observing the present ideologies in the informational context. The thematic axis of informational competence for learning and education is centered on lifelong learning, addressing the habits of inquiry in formal and informal educational practices regardless of levels and ages both in the community and in corporate environments.

6 Conclusion

The immersion of people in the digital world is anticipated by the ease and use of information and communication technologies. And the fact of having access to technological resources demands the creation of new forms of learning and digital literacy, presented as the premise of the moment. In addition, other coping measures can be taken to make people aware and lead them to understand that not all content to which they have access through social networks and media is suitable to inform. The emergence at all times of new technologies that allow unrestricted access to any and all news published on social networks and media reminds us that it is also necessary to provide security to those who search for news. They need instruments and knowledge to identify whether the information comes from a safe and reliable source.

References

Andrews, E.: The Great Moon Hoax, The History Channel. Retrieved from: http://www.history.com/news/the-great-moon-hoax-180-years-ago?linkId=16545579 (2018)

Bentes, I.: A cultura da desinformação na era digital. In: Seminário Jornalismo: as novas configurações do quarto poder. São Paulo, SESC Vila Mariana, agosto 2018. Retrieved from: https://www.sescsp.org.br/online/artigo/12366_A+CULTURA+DA+DESINFORMACAO+NA+ERA+DIGITAL (2018)

Belluzzo, R.C.B., Simeão, E.L.M.S., Santos, R.B.: Competências na era digital: desafios tangíveis para bibliotecários e educadores, vol. 6, no. 2, pp. 30–50. ETD—Educação Temática Digital, Campinas (2005)

Belluzzo, R.C.B., Simeão, E.L.M.S., Santos, R.B.: Competência em informação (CoInfo) no bibliotecário protagonista: estudo do perfil da Rede de Bibliotecas de Pesquisa do MCTIC à luz do Diagrama Belluzzo®1. Inc. Soc., Brasília, DF, vol. 8 no. 1, pp. 89–100 (2014)

Capurro, R.: Foundations of information science: review and perspectives. In: International Conference on Conceptions of Library and Information Science, 1991, Tampere. Electronic Proceedings... University of Tampere, Tampere. Retrieved from: http://www.capurro.de/tampere91.htm (1991)

Capurro, R. et al.: O conceito de informação. Perspectivas em Ciência da Informação, [S.l.], vol. 12, no. 1. ISSN 19815344. Retrieved from: http://portaldeperiodicos.eci.ufmg.br/index.php/pci/article/view/54 (2007)

Coiro, J., Knobel, M., Lankshear, C., Leu, D.J.: Handbook of Research on New Literacies, 1st edn. Routledge, New York (2008)

Darnton, R.: A verdadeira história das notícias falsas. Retrieved from: https://brasil.elpais.com/brasil/2017/04/28/cultura/1493389536_863123.html (2017)

Dudziak, E.A.: Os faróis da sociedade de informação: uma análise crítica sobre a situação da competência em informação no Brasil. Informação & Sociedade. João Pessoa, vol. 18, no. 2, pp. 41–53 (2008)

Herzstein, R.: The Most Infamous Propaganda Campaign in History, p. 492. GP Putnam & Sons. (NY) (1978). See also: Kallis, A.: Nazi Propaganda and The Second World War, p. 6. Palgrave Macmillan, New York (2005)

Lee, G.: Q&A on Cambridge Analytica: The Allegations So Far, Explained, FactCheck, Channel 4 News. Retrieved from: https://www.channel4.com/news/factcheck/cambridge-analytica-the-allegations-so-far (2018)

Leu, D.J., Forzani, E., Rhoads, C., Maykel, C., Kennedy, C., Timbrell, N.: The New Literacies of Online Research and Comprehension: Rethinking the Reading Achievement Gap. Read Res. Q. 50(1), 37–59 (2014)

McIntyre, Lee: Post-Truth: The MIT Press Essential Knowledge series. MIT Press, Cambridge, MA (2018)

Metz, R.: CNN Business. Lawmakers warn of 'deepfake' videos ahead of 2020 election—CNN. Retrieved from: https://edition.cnn.com/2019/06/12/tech/deepfake-2020-detection/index.html (2019)

Miller, J.: A Nation Challenged: Secret Sites; Iraqi Tells of Renovations at Sites For Chemical and Nuclear Arms (2001). The New York Times. See also: Miller, J.: After Effects: Prohibited Weapons; Illicit Arms Kept Till Eve of War, An Iraqi Scientist Is Said to Assert. The New

York Times. Retrieved from: http://www.nytimes.com/2003/04/21/world/aftereffects-prohibited-weapons-illicit-arms-kept-till-eve-war-iraqi-scientist.html (2003)

Neisser, F.G.: Fact-checking e o controle da propaganda eleitoral. Revista Ballot 1(2), 178–212 (2015)

Pardo. T.: Ferramenta para detectar fake news é desenvolvida pela USP e pela UFSCar. Retrieved from: https://www.icmc.usp.br/noticias/3956-ferramenta-para-detectar-fake-news-e-desen-volvida-pela-usp-e-pela-ufscar (2018)

Posetti, J., Matthews, A.: A short guide to the history of 'fake news' and disinformation. Int. Center Journalists (UCFJ) (2018)

Rabaglio, M.O.: Seleção por competências. Educator, São Paulo (2001)

Rafael, A., et al.: Contributions to the Study of Fake News in Portuguese: New Corpus and Automatic Detection Results. 13ª Conferência Internacional de Processamento Computacional do Português. PROPOR (2018)

Rehm, G.: Language Technologies for Multilingual Europe: Towards a Human Language Project. Strategic Research and Innovation Agenda. CRACKER and Cracking the Language Barrier federation, December 2017. Version 1.0. Unveiled at META-FORUM 2017 in Brussels, Belgium, on November 13/14. Prepared by the Cracking the Language Barrier federation, supported by the EU-funded project CRACKER (2017)

Saracevic, T.: Ciência da Informação: origem, evolução e relações, vol. 1, no. 1, pp. 41–62. Perspec. Ci. Inf., Belo Horizonte (1996)

Sateli, B., Cook, G., Witte, R.: Smarter Mobile Apps through Integrated Natural Language Processing Services. Semantic Software Lab, Department of Computer Science and Software Engineering, Concordia University, Montreal, Canada iLanguage Lab, Montreal, Canada (2017)

Schrader, A.M.: The domain of information science: problems in conceptualization and in consensus building. Inf. Serv. Use 6, 169–205 (1986)

Simeão, E.L.S., Marques, M., Cuevas, A.C.: Mediação e ação comunicativa: conformando nuvens e formando competências para a mediação nas redes sociais virtualizadas. Ciência da Informação (Online) 43, 241–256 (2016)

Subramanian, S.: Inside the Macedonian Fake News Complex. Wired. Retrieved from: https://www.wired.com/2017/02/veles-macedonia-fake-news/ (2017)

The New York Times: From the Editors: The Times and Iraq. Retrieved from: http://www.nytimes.com/2004/05/26/world/from-the-editors-the-times-and-iraq.html (2004)

Unesco: Journalism, 'Fake News' & Disinformation: Handbook for Journalism Education and Training. Organização das Nações Unidas para a Educação, a Ciência e a Cultura (Unesco) 2018

Wilson, C.: Alfabetização midiática e informacional: currículo para formação de professores, p. 194. Unesco, UFTM, Brasília (2013)

Gislane Pereira Santana Professor in Computer Science at Centro Universitário de Brasília Uniceub and Centro Universitário UNIEURO. Founder of the Digital Inclusion initiative, Practical Project, which teaches computer science to the poor community of Vila Estrutural in Brasília. Coordinator of the Young Coders project at Instituto Illuminante, which teaches computer programming to public school students in DF. Director of social transformation at the Illuminante Institute for technological innovation and social impact. Information Technology Project Manager and Information Systems Coordinator. Bachelor in International Relations, M.Sc. in Knowledge Management and Information Technology by the Catholic University of Brasília. PhD student in Information Science at the University of Brasília—UNB. It studies the subjects related to misinformation e Social Networks, Digital Divide (digital inclusion) as a means for social insertion.

Elmira Luzia Melo Soares Simeão Professor at the University of Brasília (2003), with post-doctorate at the Complutense University of Madrid (UCM-2019), Ph.D. in Information Science from the University of Brasília (2003), Master in Communication and Culture from the Federal University of Rio de Janeiro (1998). She has a degree in Social Communication from the Federal University of Piauí (1990). She is a full professor at the Faculty of Information Science, of the Undergraduate Course in Library Science, and researcher in the Information Science Research Program. Member of the scientific committee of the Ibero-American Journal of Information Science (international journal under the editorial responsibility of the Faculty of Information Science). She works in the area of editing, collections and skills-CoInfo. From 2010 to 2018/01, she was director of the Faculty of Information Science at the University of Brasilia, being its first director. She is a representative of the University of Brasilia in partnership with the Complutense University of Madrid (UCM, Spain), where she maintains contact with researchers from the UCM Department of Library, Documentation and Information Science. She is the leader of the research groups "Extensive Communication and AV3" and "Competence in Information and Vulnerable Populations" certified by the National Research Council of the Ministry of Science, Technology and Innovation of Brazil.

Semantic CMS for Newsrooms

Proposal of Virus Ontology to Enhance Knowledge Representation in Authoring Environments

Edgard Costa Oliveira, Edison Ishikawa, Vitor Silva de Deus, Lucas Hiroshi Horinouchi, and George Ghinea

Abstract Journalists are increasingly under pressure to produce more news in less time. In this context, a semantic CMS is a tool that helps them to search for texts related to the subject they are writing at the same time as making semantic annotation of what they produce. This work describes the creation of the Zika ontology by the MDM project team, a new issue at the time of the epidemic, for annotating news about Zika, as well as the specification, modeling and development of a semantic CMS that uses this ontology for the semantic representation of texts. This work shows the proposal of the use of ontologies during the moment of the text production, unlike other approaches in which ontologies are used mainly for post-publication of the news or for information retrieval.

Keywords Newsroom · Sematic web · CMS · Ontology · Zika

E. C. Oliveira (✉) · E. Ishikawa · V. S. de Deus · L. H. Horinouchi
Universidade de Brasília, Campus Darcy Ribeiro, Brasília 70910-000, Brazil
e-mail: ecosta@unb.br

E. Ishikawa
e-mail: ishikawa@unb.br

V. S. de Deus
e-mail: vitor.deus@alana.ai

L. H. Horinouchi
e-mail: lucas.hh@aluno.unb.br

G. Ghinea
Brunel University London, Uxbridge UB8 3PH, UK
e-mail: george.ghinea@brunel.ac.uk

© The Author(s), under exclusive license to Springer Nature Switzerland AG 2022   139
B. Medeiros Neto et al. (eds.), *Digital Convergence in Contemporary Newsrooms*,
Studies in Systems, Decision and Control 370,
https://doi.org/10.1007/978-3-030-74428-1_9

1   Introduction

Text authoring can be seen as a similar practice to those taken four centuries ago (Johann Carolus published the first recognized newspaper in Strasbourg 1605 (Weber 1605, p. 387–412), 166 years after the press invention by Gutenberg in the same city), with a slight difference: we have shifted from the hand-pen-paper model in cellulose (that still exists), to the digital finger-keyboard-cursor-white page. In the support level, a lot has changed—such as making links to other documents; making and sending as many digital copies as desired—as we can see from the development of editing resources, which were in the past restricted to editing houses and their printing press machines. In the syntactic level, we can benefit from searching and ordering key words. However, in the semantic level, text production is the same as before: it depends on the writer's ability to associate his contents to existing formal concepts structures (links to other documents and web pages, by associating text to dictionaries, terminologies, taxonomies, indexes etc.).

In the Semantic Web, we are facing a new opportunity to use concept referencing in text—and not only its objects and components such as summaries, images, links, descriptive terms and their meanings. The main problem we are facing nowadays is that the available content on the Web is generated by one person, indexed by another and retrieved by computers that do not make a difference between variant terms.

Based on previous studies Oliveira et al. (2004), we have defined an ontology-based authoring environment for the Semantic Web as "a set of writing tools for writing, editing and representing documents that interactively support users (authors), allowing a better access and use of knowledge semantic representation during writing, by doing the following tasks: semantic annotation of documents, metadata creation, linking terms in the document with external ontologies; linking similar documents with each other, transforming citations in labelled links, etc."

Particularly, when it comes to preparing a journalistic text, users of CMSs— Content Management Systems—in newsrooms, they count with a blank screen to insert texts with basic formatting options that current editors offer. However, the problem is that these tools limit the use of correct terms, by not giving the author the awareness of using the best term to identify a certain subject as well as its semantic or syntactic variations. To identify the best keywords to label the subject, to produce tags that are semantically linked, other than hanging loose and ambiguous. This happens to be the case of the subject Zika, disease or virus. The impact of this problem is related to the news production: they may contain useful information, but they were not well represented via keywords or hash tags. At the moment of this research Zika was a trend topic at newspapers in Brazil and the world. This new theme could expose the difficulties journalism and journalist will face when using semantic CMS tools in order to develop better semantic tools for newsrooms. Our research question is: can we propose an ontology-based CMS for the production of news articles that is able to link a term with its semantically related classes or instances in the ontology?

This chapter describes the conceptual and practical basis for the specification, modelling and design of an ontology-based news authoring environment for the Semantic Web, that takes into account the construction and use of an ontology of the Zika disease. It has been said that CMSs are being adapted in order to receive semantic features, such as automatic generations of keywords, semantic annotation and tagging, content reviewing etc. We present here the infrastructure designed to foster research on semantic CMSs as well as semantic web technologies that can be integrated into an ontology-based news authoring environment and at the end, as future works, the challenges we have to overcome to use these tools in a real newsroom context.

2 Methodologies of This Proposal

In this section we intend to present the context of the creation of the solution, its motivation and proposals, by indicating a semantically computational platform designed to receive the solution; the Zika Ontology construction process; a general model of the architecture and the support of the authoring environment via a semantic CMS.

This research started with a general specification of the environment, by using as a starting point the general requirement of ontology based authoring environments Oliveira (2006). In this project, we had the collaboration of undergraduate and master students from the University of Brasília (UnB), and professors from Information Science, Computer Science and Software Engineering. The group worked under the program of the Advanced Topics Computer Course. We invited a group of local medical staff to join in the construction of a Zika ontology. The users of the solution are journalists from the Campus Online UnB's newspaper, from the Communication Faculty. We divided the group in three parts: environments specification team; conceptual modelling and ontology construction team; requirements specification and software development process and engineering team.

The proposal of this tool is to annotate terms and concepts used by writers/journalists and to relate these terms with the subject ontology, to create links with other information resources about that subject: existing news pages or any other page selected by the writer. Regarding the user's side, text can be produced by many journalists and go under different review processes. However, tagging and terminology consistency will be provided and supported by the ontology that will guide users in the task of choosing a term to be used, and thus by making the links between this term and its related concepts (synonyms, hyperonyms, antonyms, related subjects etc.).

In order to implement the project, we found two viable paths: one using Python programming language (Python 1991) and another one based in the Java platform (Apache Jena 2011). Both did meet the project's requirements, so initially we opted to use a Python base framework that students can learn and prototype the tools and a Java implemented semantic database, the Jena-Fuseki. The Python platform used as the front end of the system is composed of machines based in an Intel X86 Architecture, Linux Ubuntu O.S., Apache Server, applications in Python and libraries in Python RDFlib for the data treatment in the RDF and Django formats for the CMS (Django 2005). At the back end, we used the Jena Semantic database to store RDF triples. Then, we continued this research and decided to follow the IKS (2013) architecture. Basically, IKS has two stacks, the conventional stack which implements the Information System (IS) and the Semantic Stack that handles all the IS semantic information. In this case, we implement both stacks using Python.

3 Description of the Representation Languages Used

Since the start of this project, due to the team know-how, we decided to use Python to manipulate RDF models. Considering that RDF was created to describe resources on the Web, the resource description framework is of great importance to help finding a way to extract relevant resources. Therefore, we have decided to used RDF/XML due to its simplicity and as a first formalized serialization, according to Oliveira (2006), as a working model to represent the base ontology, initially constructed in OWL format. We intend to browse the structure of the ontology and to recover classes and instances which were more relevant to the specific context. We also identified the relations between classes and instances that were listed in the ontology. Python (1991) with RDFlib (2009) worked well for the initial development of the application. Schiessl (2015) reveals that the RDFlib library is easy to use via parsers and serializers of RDF/XML data and is best used in small projects. We proposed, thus, to use RDFlib to deal with RDF and OWL data in a Python environment and SPARQL implementations.

Python is a small-scale language (Gandon and Schreiber 2014) recommended for optimized performance and has simple implementation characteristics. We learned that Apache Jena (2011) Fuseki version 1.1.0 has better performance than RDFlib. The free and open source solution based in Java is largely used to the development of Semantic Web applications. It showed efficiency in storing RDF data, good interface to submit Sparql queries and good answer performances. Even though it does not use a specific API to connect to the Jena-Fuseki server, the problem was solved by using commands via operational system to reach the goals. We used Apache Jena, a large-scale Java platform, designed for optimized performance. Indexation takes place via semantic annotation. This process is necessary to unite and interlink documents in a semantic space defined by the domain ontology. NLP —Natural Language Processing—is mainly used to identify, compare and annotate documents, however, searching for minimizing possible effects of ambiguity, NLP was complemented by human validation. Due to the heterogeneity of this platform, Python plus Java and a SGBD were used both for implementing the conventional stack and the semantic stack, by making the implementation complex, and the projects tema learning curve was longer, thus leaving almost no time to play with other implementing semantic tools, which we believe were the funniest part. Implementing everything with Python makes prototyping semantic tools easier. Thus, we further show the initial Python/Java architecture implementation and later our implementation of the IKS (2013) architecture with Python only.

The semantic annotation steps are as follows (RDFLIB 2009):

(a) extract all ontological entities and lexical variations to a list;
(b) analyze documents and remove symbols and non-relevant text;
(c) analyze the text in order to extract relevant terms and lexeme;
(d) identify n-grams or other patterns;
(e) eliminate stop words;
(f) compare with the ontology labels;
(g) indicate a grammar class to the term;
(h) indicate similarity of the term with the domain meaning;
(i) confirm the annotation via a domain specialist; and
(j) add the annotation to the documents.

The RDFS–RDF Schema (W3C 2014) is a semantic extension of RDF and offers mechanisms to describe related groups of resources and the relationships between them. Daconta et al. (2003) present the main components of this vocabulary, described as follows: Classes:rdfs:Resources—it is the class of all other classes which are subclasses of this one; rdfs:Class—defines a group of related entities that share the same properties; rdfs:Literal—represents constant values such as texts and numbers; rdf:Property—defines a property of a class and the representing value; rdfs:domain—defines which class of a property it belongs to; rdfs:range—defines a group of possible values to a property; rdf:type—a standard property to define an RDF subject in an RDF Schema; rdfs:subClassOf—specifies that a class is a specialization of another one; rdfs:subPropertyOf—declares that all resources that are related by a property are also related to other ones; rdfs:label—is an attribute that defines a label that is readable by humans.

To perform the search in the database, we used Sparql, which is a standard search language and a data access protocol. It means that Sparql allows more than access to the RDF triples—subject-predicate-object—or graphs but also to any data sources that can be mapped in the RDF. When using the Python semantic stack instead of Java/Fuseki, we used Sparql and algorithms coded in Python that made the system more flexible.

4 Ontology Construction Methods Used

During the setting up of the authoring environment, a specific project-working group responsible for the ontology had meetings with a team of medical staff from the Health Department of the Government of Brasília/DF, in order to generate a conceptual map of the Zika Disease (Fig. 1). The meetings lasted 10 h total approximately and the medical team informed all their knowledge about the virus and the context of the disease. The focus of the Zika conceptual map was to annotate news articles about Zika with the help of medical specialists. The first artifact produced was the conceptual map for the understanding of the ontology

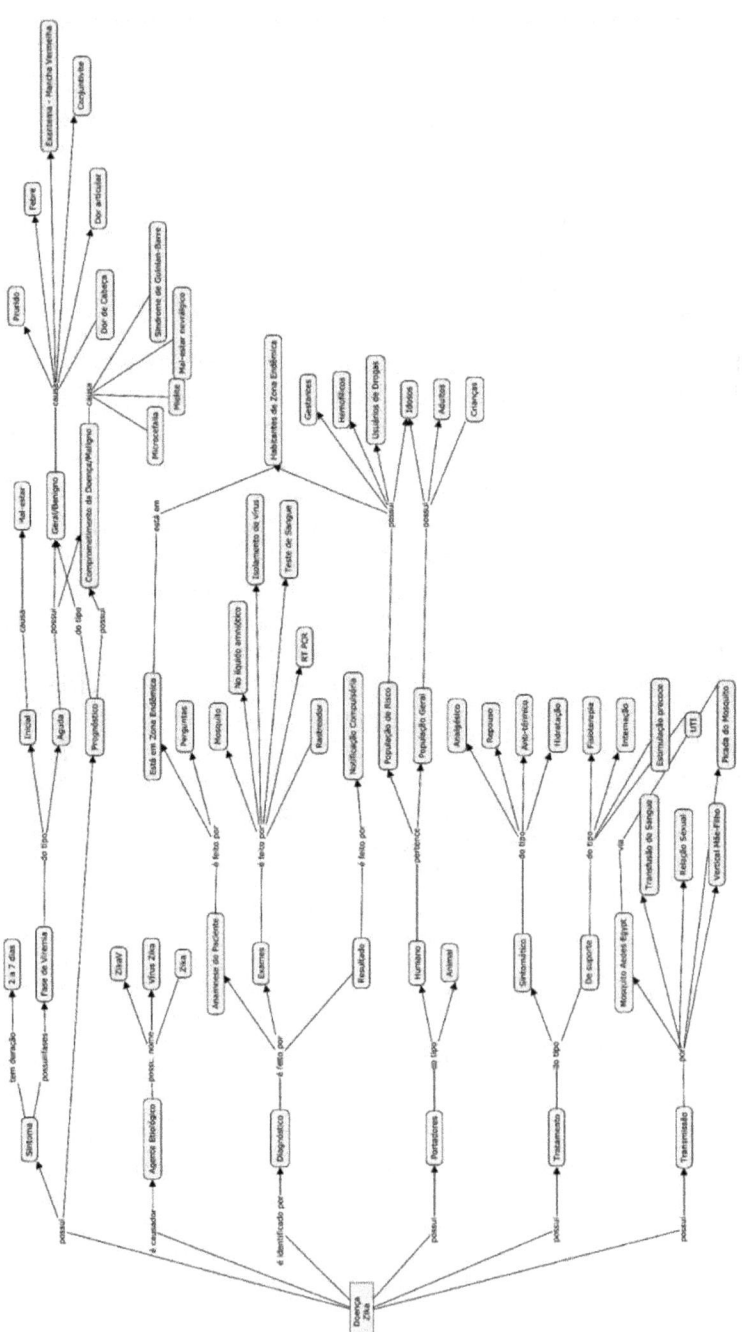

Fig. 1 Zika conceptual map (in Portuguese). *Source* Authors (2016). Free translation

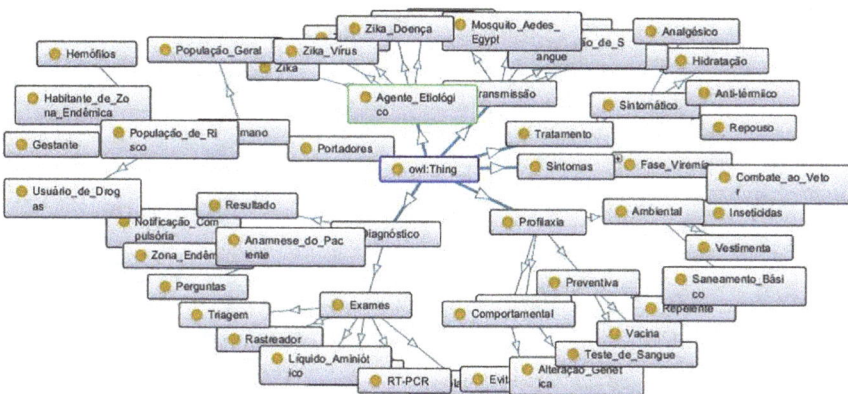

Fig. 2 The Zika ontology constructed by the project's team. *Source* Authors (2016). Free translation

domain. The teams homologated the results after the edition, and there were some adjustments necessary for the construction of the ontology.

We used the 101 Methodology (Noy and McGuinness 2001) to create the Zika ontology, from the University of Stanford, a simple method, whose authors also developed the ontology environment such as Protégé, Ontolingua e Chimaera (Isotani and Bittencourt 2015). This method is divided in phases: (1) Scope definition—from the meetings with the medical group, we defined the scope of the ontology.; (2) Consider reuse—there was no other ontology specifically about Zika, but some had Zika as an instance, however we based our research on the structure of these ontologies (CRRD 2016; RGD 2016; BioportalSnomedCT 2016) and resource documents (Schram 2016; Bushak 2016; Rasmussen et al. 2016); (3) Enumerate terms—all terms were numbered via XMind and then via CMapTools, from the meetings with medical staff as well as from searching the theme in specialized medical bibliography and news articles; (4) Define classes—this was complex and divergent, because deciding what is class or subclass can be confusing and time consuming. (5) Define properties—each class properties were identified in the conceptual map and were simple to implement in the ontology; (6) Define restrictions—they were not used at this time due to scope and time limitations; (7) Create instances—after we reviewed all classes and properties, we defined which were to become instances of the Zika ontology.

When considering reuse in the 101 Methodology, we identified that the term Zika is considered an instance within the ontology of diseases—Diseases (RDO:0000001) (CRRD 2016) and Zika Virus Infection (RDO:0016040) (RGD RGD 2016)—as presented in Fig. 3a, b, respectively.

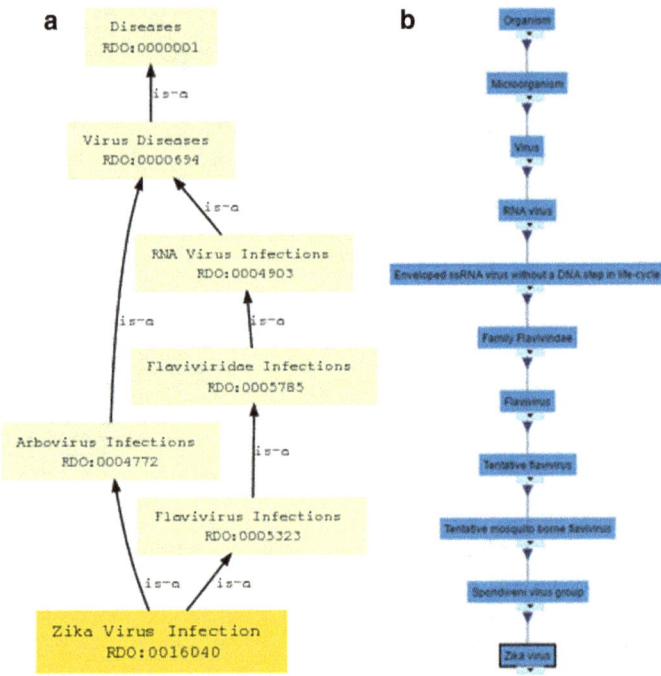

Fig. 3 **a** Zika as an instance of disease ontology. **b** Zika as an instance of virus ontology. *Source* Authors (2016). Free translation

Finally, the ontology was constructed by using the Web Protegé.[1] This ontology was then validated via Hermit reasoner and by the group responsible for its construction, including the medical staff and other specialists in the subject area and can be seen in Fig. 2.

5 Proposed Solution Design

We searched for recommendations from the W3C about ontology driven applications as well as the ontology-based authoring environment previously designed Oliveira (2006). Some of our main questions surrounded the following issues: RDF is a kind of a set of individual data saved in a schema based on an OWL graph. The retrieval of this data is made via Sparql, however there are some limitations: How can we write an RDF at each new register of an application? How can we modify

---

[1]Available at the project's ontology page: http://webprotege.stanford.edu/#Edit:projectId= 7515ad86-1bdc-431f-a7a3-b9b8167ec068.

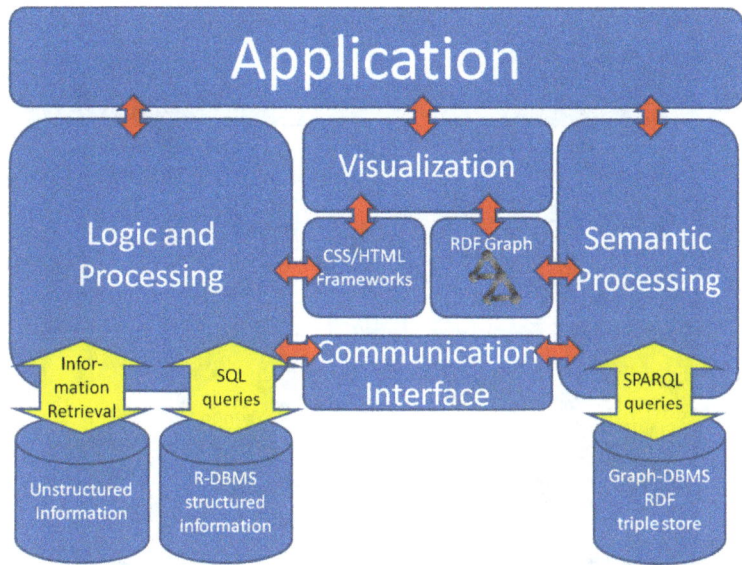

Fig. 4 Model using the application via structural databases and ontologies. *Source* Authors (2017). Free translation

this same RDF? Will it be necessary to create and RDF every time the new register is filled in? Is it possible to convert OWL to a structure model in SQL?

Then we defined some scenarios: today´s use of structural databases, as showed in Fig. 4 is a structured application that works in 3 main areas: logic and processing, databases and visualization. The user visualizes the screens, makes requests to the program, and processes the requests for data from the base, which returns the data that is interpreted by the application and shown to the user.

Figure 4 presents a general view of the system architecture that uses a series of different architectural views to illustrate the different aspects of the system. The intention is to capture and transmit the main decisions that were taken in regards to the system, from an architectural point of view. We evolved this architecture with the left side in python and the right side in Java to our view of IKS via a Python implementation (Fig. 5), presented further in item 5.1 of this article.

6 Prototype Proposal of the Ontology-Based Authoring Environment

In a simplified manner, we present here how the project implementation was modeled, by showing each tool and language used. In order to implement the project, we initially found two viable paths: one using Python, according to the

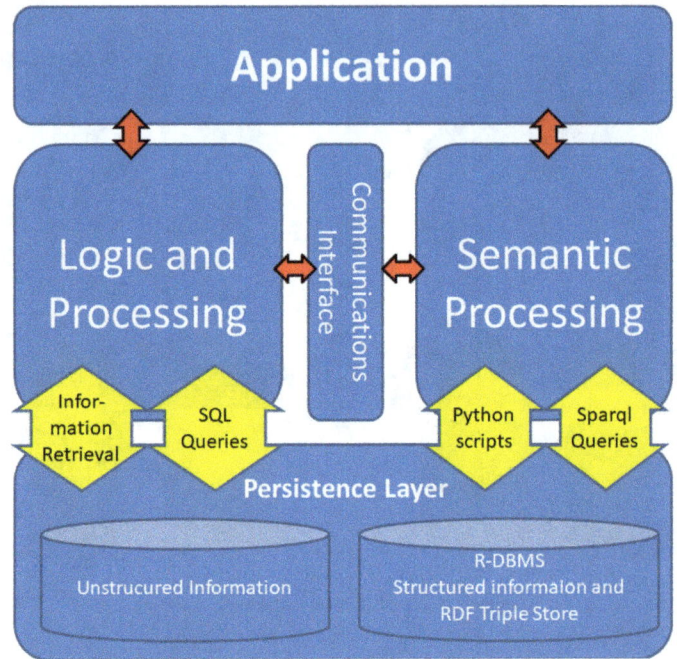

Fig. 5 Evolved IKS architecture via Python. *Source* Authors (2017). Free translation

views of Talas et al. (2011) and another one based in the Java platform, as suggested by Isotani and Bittencourt (2015). This research was conducted in an undergraduate course research, so we were free to try both paths.

The Python platform is composed of machines based in an Intel X86 Architecture, Linux Ubuntu O.S., Apache Server, applications in Python and libraries in Python RDFlib for the data treatment in the RDF and Django formats for the CMS.

Thus, the text authoring environment interface can present 3 modules: writing, ontology and semantic search engine:

1. we use Django to present a window of the document being edited;
2. another one with an RDF graph, corresponding to the semantic document annotation of the text being edited; and in a
3. third window of the returned documents from the semantic search engine, that also uses the RDF graph of the document to search for semantically related document.

The Java platform was set up with the following specifications. In a machine with Ubuntu O.S., we installed the Apache Tomcat software, so that it is possible to manage a local sever based on Java servlets and supported by a Semantic CMS, thus suggested the Apache Stanbol (2016). In the issue of programming languages and supporting applications in the treatment of RDF files, we proposed the use of

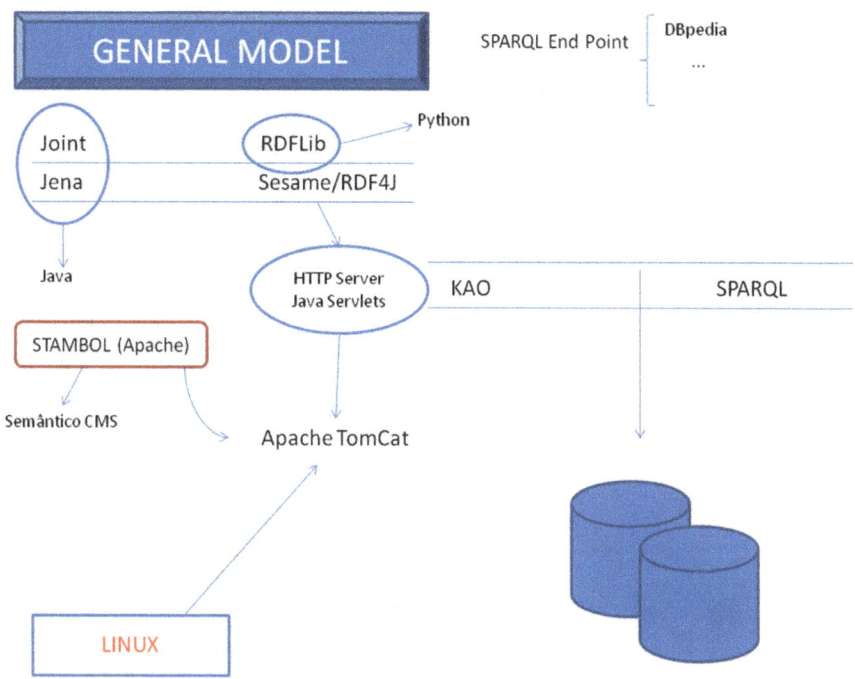

Fig. 6 General view of the model's components. *Source* Authors (2017). Free translation

the applications Joint and Jena in Java, also counting on the support from Sesame/RDF4J for handling of RDF files. Finally, information search and retrieval are based on Sparql, and this standard language used in semantic applications can be supported by the KAO implemented by Joint in order to refine the searches and retrievals (Fig. 6).

Figure 7 shows a summary of the solution components model, where the application database comes from the defined domain ontology. While editing a text in a wiki environment (i.e. MediaWiki), the platform recognizes the text edited via annotation with a Sesame RDF4J Joint and Jena application, and via servlets Java. Then it interacts with the Apache Tomcat and Apache Stanbol (2016), which is the solution for a semantic CMS. Classes and instances of the ontology are then matched with the text in order to support the annotation process. Considering the document treatment process, an XML Zika-subject text is being written, while annotations are made via connection with the OWL ontology, then making the semantic annotations and thus extracting a sub-graph with the ontology instances or classes that were recognized in the text. This sub-graph or sub-ontology is used to generate specific and context aware keywords and tags to represent the text, as well as to hyperlink it with other strictly related texts that are similar in concepts and terms.

When we restricted the environment to a Python-only implementation, the resulting system is much simpler and easier to learn (Fig. 8).

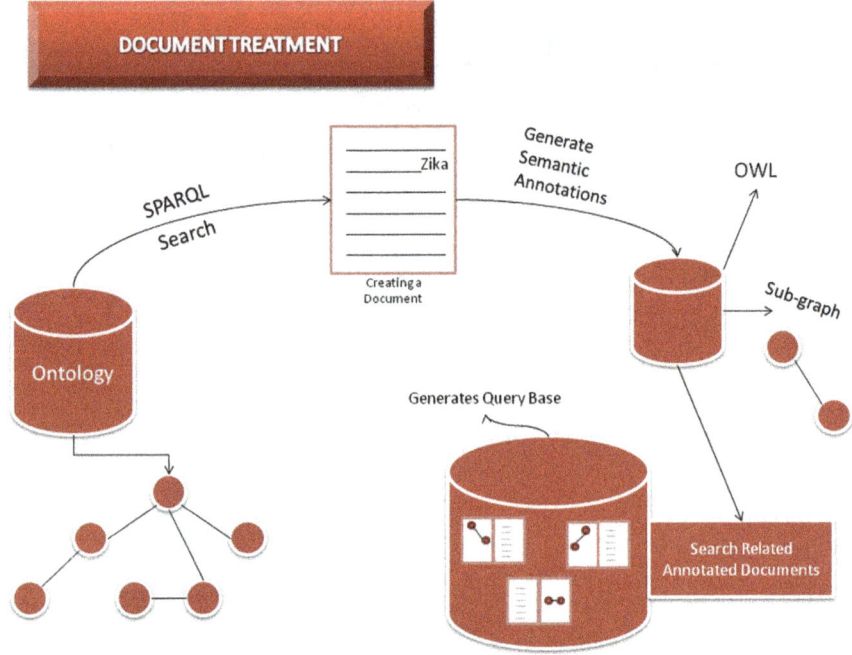

Fig. 7 Document treatment view of the system architecture. *Source* Authors (2017). Free translation

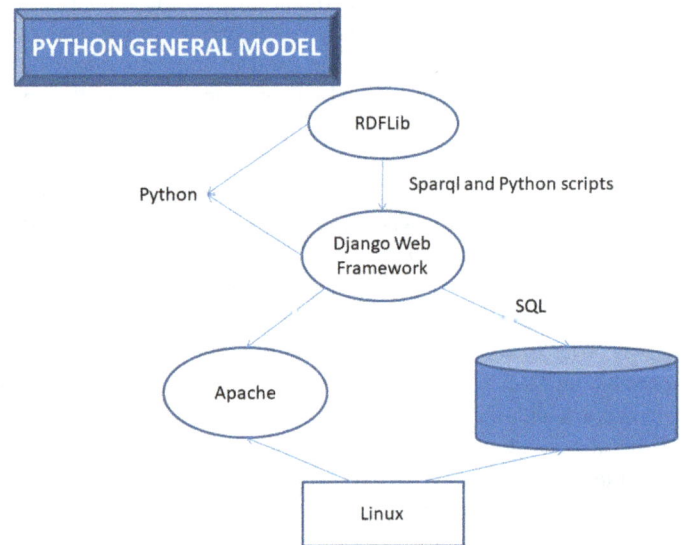

Fig. 8 Python-based general model. *Source* Authors (2017). Free translation

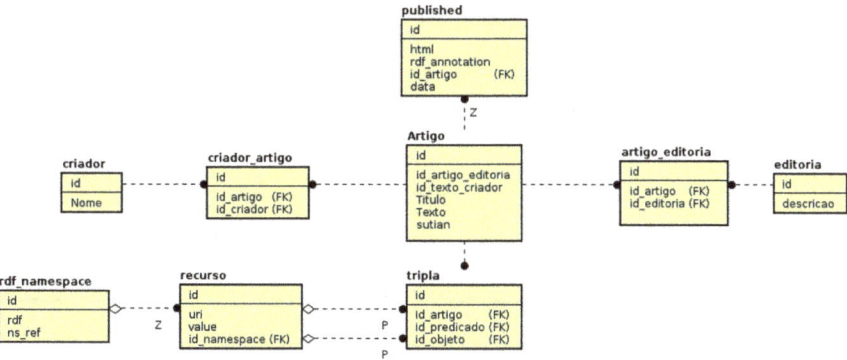

Fig. 9 Database model for a Semantic CMS. *Source* Authors (2018). Free translation

Django Framework uses the Model-View-Controller (MVC) design pattern, and we have used this model in Fig. 9 to show the database model for our CMS (de Deus 2018). This model shows that an article consists of a title, text, subtitle, editorials and authors. Each article can be related to several editorials and several authors as well as a given editorial and a given author can be related to several articles. Each article has zero or more related published articles. Published articles have the content of the article that appears on the web (HTML), the RDF file semantic annotations and a publication date. Each article is related to zero or more triples that are constructed from the semantic annotation of the article text. The triple is represented by a table in the relational bank and contains a reference to its article (subject) reference to a resource (predicate) and a resource (object). Each resource has a URI and a value, which are the URI and the Ethiological Agent. Therefore, the URI field is the replication of a Semantic Web identifier in the prototype's internal relational database.

This data base model was inspired by the solutions found in McBride (2000), which present solutions for storing RDF triples in relational banks.

7   The Semantic CMS for Newsrooms

The resulting semantic CMS (de Deus 2018) was customized to be used in newsrooms by journalists. To simulate the semantic CMS, we use the previous Zika ontology built specifically for text annotation. The screen for creating and editing articles is shown in Fig. 10. The title fields, text and subtitle are textual insertion while editorial and authors are multiple selection.

The "Annotate" button executes the text annotation algorithm and saves the article generating a list of concepts found according to Fig. 11.

Concepts can be cleared from the list by the author who is taking notes, if they have not in fact a semantic match with the text, and they can also be added via the

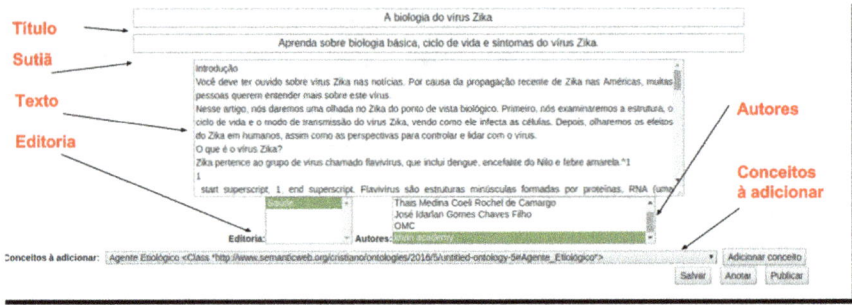

Fig. 10 CMS interface for creating and editing articles. *Source* Authors (2018). Free translation

Conceitos presentes no texto

- ☑ http://www.semanticweb.org/cristiano/ontologies/2016/5/untitled-ontology-5#Agente_Etiológico - Agente Etiológico
- ☑ http://www.semanticweb.org/cristiano/ontologies/2016/5/untitled-ontology-5#Aguda - Aguda
- ☑ http://www.semanticweb.org/cristiano/ontologies/2016/5/untitled-ontology-5#Benigno - Benigno
- ☑ http://www.semanticweb.org/cristiano/ontologies/2016/5/untitled-ontology-5#Comprometimento_da_Doença - Comprometimento da Doença
- ☑ http://www.semanticweb.org/cristiano/ontologies/2016/5/untitled-ontology-5#Diagnóstico - Diagnóstico
- ☑ http://www.semanticweb.org/cristiano/ontologies/2016/5/untitled-ontology-5#Exames - Exames
- ☑ http://www.semanticweb.org/cristiano/ontologies/2016/5/untitled-ontology-5#Fase_Viremia - Fase Viremia
- ☑ http://www.semanticweb.org/cristiano/ontologies/2016/5/untitled-ontology-5#Febre - Febre
- ☑ http://www.semanticweb.org/cristiano/ontologies/2016/5/untitled-ontology-5#Gestante - Gestante
- ☑ http://www.semanticweb.org/cristiano/ontologies/2016/5/untitled-ontology-5#Humano - Humano
- ☑ http://www.semanticweb.org/cristiano/ontologies/2016/5/untitled-ontology-5#Microcefalia - Microcefalia
- ☑ http://www.semanticweb.org/cristiano/ontologies/2016/5/untitled-ontology-5#Mosquito - Mosquito
- ☑ http://www.semanticweb.org/cristiano/ontologies/2016/5/untitled-ontology-5#População_de_Risco - População de Risco
- ☑ http://www.semanticweb.org/cristiano/ontologies/2016/5/untitled-ontology-5#Portadores - Portadores
- ☑ http://www.semanticweb.org/cristiano/ontologies/2016/5/untitled-ontology-5#Preventiva - Preventiva
- ☑ http://www.semanticweb.org/cristiano/ontologies/2016/5/untitled-ontology-5#Profilaxia - Profilaxia
- ☑ http://www.semanticweb.org/cristiano/ontologies/2016/5/untitled-ontology-5#Repelente - Repelente
- ☑ http://www.semanticweb.org/cristiano/ontologies/2016/5/untitled-ontology-5#Sintomas - Sintomas
- ☑ http://www.semanticweb.org/cristiano/ontologies/2016/5/untitled-ontology-5#Transfusão_de_Sangue - Transfusão de Sangue
- ☑ http://www.semanticweb.org/cristiano/ontologies/2016/5/untitled-ontology-5#Transmissão - Transmissão
- ☑ http://www.semanticweb.org/cristiano/ontologies/2016/5/untitled-ontology-5#Zika - Zika

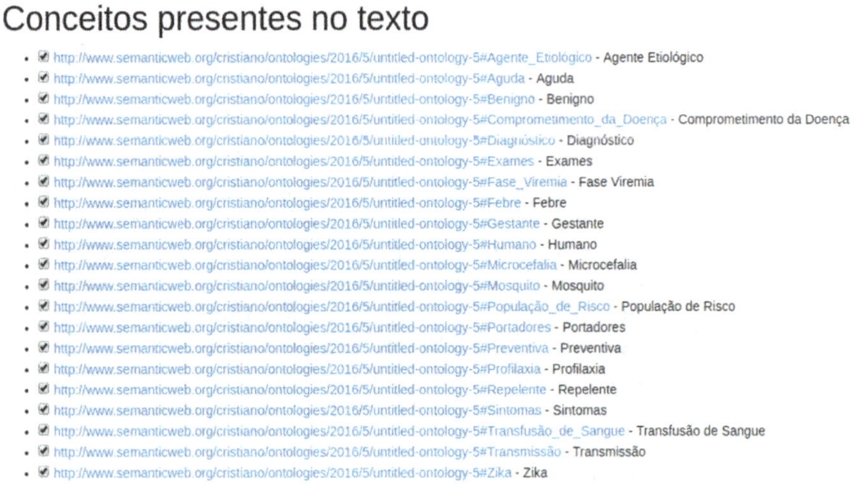

Fig. 11 Results of annotating an article about Zika

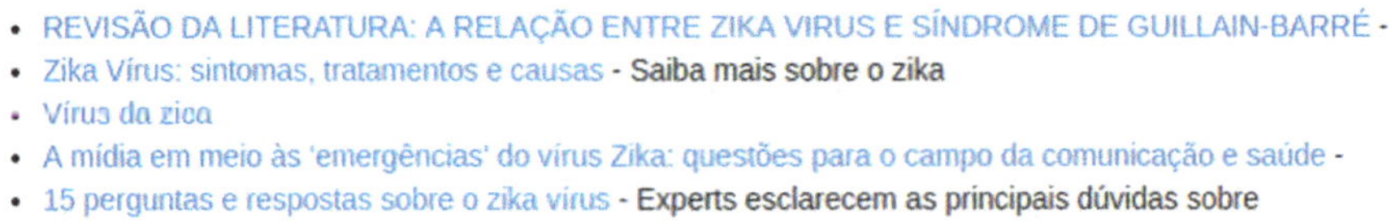

- REVISÃO DA LITERATURA: A RELAÇÃO ENTRE ZIKA VIRUS E SÍNDROME DE GUILLAIN-BARRÉ -
- Zika Vírus: sintomas, tratamentos e causas - Saiba mais sobre o zika
- Vírus da zica
- A mídia em meio às 'emergências' do vírus Zika: questões para o campo da comunicação e saúde -
- 15 perguntas e respostas sobre o zika vírus - Experts esclarecem as principais dúvidas sobre

Fig. 12 Resulting list of texts related to the Zika article. *Source* Authors (2018). Free translation

selector with the label "Concepts to add". After the annotation is generated in the same page, the system shows a list of five articles allegedly most related to the article that is being edited (Fig. 12).

8 Conclusions

The main results of the experience, related in the paper, are the construction of an OWL Zika Ontology, the modeling of the authoring environment and the implementation of the database search mechanism. We presented in this chapter the experience taken in the Laboratory of Special Projects with students of the Department of Computer Science and Software Engineering of the University of Brasília with the challenge to understand and apply Semantic Web technologies to enhance the semantic capacity of CMSs.

We showed here the following contributions to the area: the specification and modeling of an authoring environment for a text editor, supported by a semantic and lexical interpreter, for the edition of news articles about Zika, supported by a specific ontology created by the project's team.

The architecture models of a prototype of a semantic CMS was described and implemented in the lab, and improved with pure Python implementation inspired in the IKS reference architecture. This model represents the effort and practice of the students who showed advanced abilities to deal with semantic web challenging issues in a computer science environment.

This experience resulted in an environment that allows the use of a text editor, integrated to a semantic CMS, in which terms can be typed and simultaneously and automatically recognized and associated to classes and instances of the Zika ontology. From the relationships created between the ontology and the text, the author is able to obtain from this annotation a list of keywords and conceptual #tags that identify specific subjects of the text, the scope of the article in relation to the general context of the Zika ontology. It also correlates the text with already existing texts and articles or pages so that they can be interconnected via non-ambiguous semantic relationships.

This work shows the feasibility in the use of ontologies during the moment of text production, that is, during the moment authors are deciding which terms to use in the text, in order to enhance information representation. The difference from other approaches is that ontologies are mostly used for post-publication or for information retrieval purposes. We also showed that it is possible to implement solutions not yet identified in existing CMSs available in the market, which have not benefited from ontology-based solutions that enhance knowledge representation capabilities.

References

W3C.: RDF Schema 1.1 Recommendation 25 February 2014 (2014). Retrieved from: http://www.w3.org/TR/rdf-schema/. Access on Apr 2020

Apache Jena.: A free and open source Java framework for building Semantic Web and Linked Data applications. Configuring Fuseki (2011). Retrieved from: http://jena.apache.org/index.html. Access on Apr 2020

Apache Stanbol (2016) Retrieved from: https://stanbol.apache.org/. Access on Apr 2020

BioportalSnomedCT.: Zika virus (2016). http://purl.bioontology.org/ontology/SNOMEDCT/50471002. Access on May 2016

Bushak, L.: A brief history of Zika Virus, from its discovery in the Zika forest to the global outbreak today (2016). http://www.medicaldaily.com/zika-virus-outbreak-history-381132. Access on Apr 2016

CRRD.: Ontology of disease. Ontology browser of the CRRD. Bioinformatics Program, HMGC at the Medical College of Wisconsin (2016). Retrieved from: http://crrd.mcw.edu/rgdweb/ontology/view.html?acc_id=RDO:0000001. Access on June 2016

Daconta, M.C., Obrst, L.J., Smith, K.T.: The Semantic web: a guide to the future of XML, web services, and knowledge management. Willey (2003)

de Deus, V.S.: Anotação semi-automática baseada em ontologia, busca e relacionamento semântico entre textos: proposta para um sistema de gerenciamento de conteúdo. Graduation monograph. Brasília, June (2018)

Django.: The web framework for perfectionists with deadlines (2005). https://www.djangoproject.com/. Access on Apr 2020

Gandon, F., Schreiber, G.: RDF 1.1 XML Syntax. Rio de Janeiro: W3C (2014). Retrieved from: http://www.w3.org/TR/rdf-syntax-grammar/. Access on May 2016

RGD.: Zikavirus infection. Gene Editing Rat Resource Center (2016). Retrieved at: http://rgd.mcw.edu/rgdweb/ontology/view.html?acc_i d=RDO:0016040. Access on June 2016

IKS.: Developing semantic CMS applications: the IKS handbook. Wernher Behrendt and Violeta Damjanovic (eds.) Salzburg Research Forschungsgesellschaftm.b.H (2013)

Isotani, S., Bittencourt, I.: Dados abertos conectados. Novatec Editora, São Paulo, Brazil (2015)

McBride, I.: Storing rdf in a relational database (2000). Retrieved at: http://infolab.stanford.edu/~melnik/rdf/db.html. Access on Apr 2020

Noy, N., McGuinness, D.: Ontology development 101: #. Stanford University (2001). http://protege.stanford.edu/publications/ontology_development/ontology101-noymcguinness.html

Oliveira, E.C., van Harmelen, F., Lima-Marques, F.: A framework for ontology-based authoring environments. In: ISWC 2004—International Semantic Web Conference, Hiroshima, Japan (2004)

Oliveira, E.C.: Autoria de documentos para a Web Semântica: um ambiente de produção de conhecimento baseado em ontologias. University of Brasília, 2006 (PhD dissertation). 260p (2006)

Python.: Python 3.x (1991). http://python.org. Access on Apr 2020

Rasmussen, A., Jamieson, D.J., Honein, M.A., Petersen, L.R (2016) Zika virus and birth defects—reviewing the evidence for causality (2016). Retrieved at: http://www.nejm.org/doi/full/10.1056/NEJMsr1604338#t = article. Access on May 2016

RDFLIB.: RDFLIB 5.0-dev (2009). Retrieved from: https://rdflib.readthedocs.io/en/latest/. Access on Apr 2020

Schiessl, M.: Lexicalização de ontologias: o relacionamento entre conteúdo e significado no contexto da recuperação da informação. 261 f., il. PhD dissertation on Information Science, University of Brasília (2015)

Schram, P.:. Zika Virus and public health. J. Hum. Growth Dev. **26**(1), 7–8 (2016). http://dx.doi.org/10.7322/jhgd.114415. Access on June 2016

Talas, J., Gregar, T., Pitner, T.. Semantically enriched tools for the knowledge society: case of project management and presentation. In: Knowledge Management, Information Systems, E-Learning, and Sustainability Research Volume 111 of the Series Communications in Computer and Information Science Springer (2011)

Weber, J.: The origins of the newspaper in Europe. German History **24**(3) (1605) (Strasburg, July 2006). https://doi.org/10.1191/0266355406gh380oa

Edgard Costa Oliveira Associate Professor at the Department of Production Engineering at the University of Brasília, EPR/FT/UnB since 2017. From 2008 to 2017 professor at the Faculty UnB Gama FGA/Software Engineering. Post-doctorate in Computer Science CIC/UnB (2017), suppervised by Brunel University London. PhD and Master in Information Science from UnB

(2006 and 2001). PhD CAPES scholarship at Vrije Universiteit Amsterdam, Department of Computer Science (2004–2005). Coordinator of the Research Line and professor of the Professional Master in Applied Computing/Governance and Risk Management at CIC/UnB. Researcher and professor in: Information Systems, Project Management, Semantic Web and Ontologies, Engineering Integrating Project, Introduction to Engineering, Governance and Risk Management, Norms and Standards for Information Security, Humanities and Citizenship for Engineering.

Edison Ishikawa He is a professor of the Department of Computer Science of the Brasília University (UnB) since 2014. He received the PhD degree in Systems and Computer Engineering from the Federal University of Rio de Janeiro (COPPE/UFRJ) in 2003, the M.Sc. degree in Informatics from Pontifical Catholic University of Rio de Janeiro (PUC-RIO), the B.E. Degree in Computer Engineering from Military Institute of Engineering and the B.S. Degree from Agulhas Negras Military Academy. His current research interests focus on Semantic Computing, Systems of Systems Engineering, Distributed Systems and Security. At UnB he participates in research projects in those areas, with funding from Brazilian government agencies and institutions such as CNPq, FAP-DF and Brazilian Army. Participated in the implementation activities of the MDM Project: A model proposal for a semantic framework of a collaborative environment for information management in journalistic writing.

Vitor Silva de Deus Bachelor of Applied Science. Skills and expertise: Software Development; C ++; SQL; C; Linux environment; Git; Parallel Programming; MPI; Shel Programming; Shel Scripting. Participated in the research project Multimodal Digital Media in Newsrooms: a semantic computational model in a convergent digital structure (*see "Towards a semantic-based content management system for journalistic writing." Available at:* http://medes.sigappfr.org/18/).

Lucas Hiroshi Horinouchi Software Engineering Student Engenharia de Software Universidade de Brasília - Faculdade UnB Gama. Intern at the Ministry of Economy. Federal Government-Brazil. Participated in the Research Project: Multimodal digital media in newsrooms: a semantic computational model in a convergent digital structure. Ontology-Based Authoring Environment Implementation Proposal for the Semantic Web.

Gheorghita (George) Ghinea is a Professor in Mulsemedia Computing in the Department of Computer Science, at Brunel University. Dr. Ghinea's research activities lie at the confluence of Computer Science, Media and Psychology. In particular, his work focuses on the area of perceptual multimedia quality and how one builds end-to-end communication systems incorporating user perceptual requirements. Dr. Ghinea has applied his expertise in areas such as eye-tracking, telemedicine, multi-modal interaction, and ubiquitous and mobile computing, leading a team of 8 researchers in these areas. He has over 300 publications in his research field and is the Editor in Chief of the International Journal of Pervasive Computing and Communications. Currently, his research pursuits are centered on extending the notion of multimedia with that of mulsemedia a term which he has put forward to denote multiple sensorial media, ie. media applications which engage three or more of the human sense. His work has been funded by both national and international funding agencies and has been covered by the BBC, Telegraph, and Forbes magazine, among others. He consults regularly for both public and private institutions in his areas of expertise.

Perspectives of the Journalists Content Production from Print Newspaper to Virtual Newsroom 4.0

Edison Ishikawa, Benedito Medeiros Neto, and Gheorghita Ghinea

Abstract News consumers are changing their behavior when accessing and interacting with news content, of which they are now *prosumers* (combined news *producers* and *consumers*). In this scenario, social media are not only a source of information, but also new channels of communication to publish customized content. Consequently, communication organizations are facing great challenges posed by the decrease of paying readers. Nonetheless, the independence of news organizations from their traditional sponsors is critical. Moreover, journalists' work conditions are deteriorating with increasing unemployment, low salaries, and heavy load of work. In this context, what we call social computing (social behavior modified by computational systems) is recreating the way news is produced and consumed. To understand the social role of the journalists and their managers in this challenge, we investigate how top news organizations are tackling this crisis. In this chapter, we transcend the results of the research, of a qualitative and exploratory nature, to evolve the framework model from Newsroom 3.0 to Newsroom 4.0, a collaborative environment to support the production of news in an integrated, convergent and cybernetic newsroom, integrating and managing social media with a tool that can be customized not only to be a cyber news platform, but also a novel social media for news.

Keywords Convergence · Cyber journalism · Semantic newsroom · 4C model · Tetrahedron framework · Scenarios and media

E. Ishikawa (✉) · B. M. Neto
Universidade de Brasília, Campus Darcy Ribeiro, Brasília, Brazil
e-mail: ishikawa@unb.br

B. M. Neto
e-mail: medeirosneto@unb.br

G. Ghinea
Brunel University London, Uxbridge UB8 3PH, UK
e-mail: george.ghinea@brunel.ac.uk

© The Author(s), under exclusive license to Springer Nature Switzerland AG 2022
B. Medeiros Neto et al. (eds.), *Digital Convergence in Contemporary Newsrooms*,
Studies in Systems, Decision and Control 370,
https://doi.org/10.1007/978-3-030-74428-1_10

1 Introduction

Change in newsrooms is nothing new. However, over the past 15 years, the pace of change has accelerated and gained a new dimension as a result of technological, cultural and economic changes). Particularly impacted has been the production process of journalistic newsrooms (Avilés 2017). These changes were summarized by Salaverría and Negredo (2008) in four central dimensions of journalistic convergence: business or economic, technological, professional and communicative. The changes in journalism are not simply driven by the strong presence of technologies, such as social network, but by the extent of knowledge about domain tasks and professional behavior (Dowd 2013).

Convergence in newsrooms has led to the presence of interdisciplinary knowledge, but it is possible to distinguish two cultures, journalistic and technological (production engineers, statisticians, design, sociologists, etc.). Certainly, there are differences, which can start with the specific interests of each of these two cultures. Journalism is concerned with the principles of objectivity, impartiality, and others associated with journalistic activity outside the newsroom. Technology and related people focus on issues related to readers' access devices, the process of producing news, and the construction of journalistic content. There are other more specific concerns, such as news recovery, multichannel distribution, and audience consumption analysis (Canavilhas 2017).

However, the goals of professionals tend to be the same, to inform the audience well. Cyberspace and new forms of sociability are largely the result of new communication devices that have the potential to transform the way man relates to himself, to his work, and to the world around him. The emergence of these new devices, of ICT (Information and Communication Technology) together with a rich variety of platforms and communication channels have laid the premises for the emergence of new behaviors, new forms of integration and new process of sociability at home, at school and at work (Temer and Nery 2009). In this context, social media is shaping the new journalism, the Social Journalism (Ermida 2012). Social Journalism is about news prosumers (combined news producers and consumers/readers) engagement, social newsgathering and verification to sustain reliable and free journalism, to overcome social media fake news (Crilley and Gillespie 2019).

Newsroom 4.0 meets the needs of Social Journalism by providing a framework model that fulfills the needs of an agile environment in the production of news stories (involving prosumers), and of quality (allowing the configuration of flexible and semantically verifiable news business processes). Moreover, although this shift of power from a traditional news media company to that of a social media one is getting journalists to use social media as news channels, in fact, they have to think about how to create new social media news using existing social media. We contend that the Newsroom 4.0 framework model can be used by journalists, journalism students and journalism professors to act as Cyber Journalists in the activity to create news social media.

The technological convergence between computing, telecommunications and the cultural industry is not an "irreversible" natural result. It is a result of the needs of economic groups that seek economies of scale in the scope of their productions and operations. As an example of Collaboration Networks and Content Exchange, the presence of social networks in the newsroom workplaces is highlighted, and in the editorial area there is the edition of e-books[1] Social networks and e-book publishing (Heller and Mello Junior 2017).

The research reported in this paper is justified for a few reasons, such as the new styles of being and acting in the interaction space of the essay, which is replete with interfaces with computers, digital networks, ubiquitous computing, hybridism, smartphone mobility and other artifacts. A second relevant justification is that, in the second decade of the twenty-first century, the techniques for the work of journalists, especially reporters and writers, diversify, and this allows for more exact texts to be written (Laje 2012). Accordingly, we adopt a technological angle and understand that convergence and integration have as a principle the collaborative coordination of individuals, and resources (Schwingel 2012).

In this context, our objective is the analysis and development of a newsroom framework model comprising the following aspects: planning, content generation and publication. The model thus elaborated is based on the visits to five big media organizations on three different continents. To this end, we review the need of convergence, semantic and integration in the newsroom of the future; describe the methodology to map the evolution of the newsroom in the context of convergence; based on this, we propose a conceptual framework to handle the contemporary and future needs of newsrooms, and exemplify its use in different scenarios; lastly, we provide suggestions for future research and practical implications of the proposed framework.

2 Convergent, Collaborative, and Semantic Newsrooms

Human development is a consequence of or is directly related to the domain of tools and their technological development. Machines, equipment or information systems, sometimes seen as platforms, allow for greater agility in the production, dissemination and access of information to a larger number of people (Temer and Nery 2009). In the context of the modern newsroom, these lead to novel convergent, semantic and collaborative dimensions, which shall now be explored in more detail.

2.1 Convergent Newsrooms

With the introduction of convergent multimedia and distinct distribution channels, newspaper companies began to require journalists to produce content for different platforms and formats (Dailey et al. 2005; Menke et al. 2016). The literature highlights presence of studies related to the analysis of the process of convergence of newspaper writing, in the following aspects: planning, content generation; and publication (Maia and Agnez 2011). The change, which occurs when a news organization had worked with traditional print media and then launches its publications on the web, has been explored by Belochio (2012), among others. Studies have also explored the implication of the distribution and expansion of *Zero Hora* contracts in the context of convergence journalism, as well as of the expansion of collaborative journalism and the use of social networks internal and external to the newsroom (Filippo et al. 2011; Rublescki and Barichello 2013).

Comparative studies, highlighting the pros and cons of convergence in Austria, Spain, and Germany have also been carried out (Avilés et al. 2009), which emphasized the need for the development of convergence and integration models. Later studies comparing editorial strategies for cross-media news production in six countries—Germany, the Netherlands, Switzerland, Austria, Spain, and Portugal—confirm the increasing trend of newsroom convergence (Larrondo et al. 2016; Menke et al. 2016).

2.2 A Model of Collaborative Environment

Journalists should be able to meet the demands of quality and time from newspaper management and at same time draft the news production to print and also to cyberspace/Internet. They are required to write reasonable texts with editorial malleability. Secondly, they should favor a permanent productive work environment in journalistic writing; thirdly, they can rely on frameworks that use collaborative technologies or groupware software (Davies 2017).

It is expected that these environments, most often composed of frameworks model and collaborative systems, facilitate communication between people, guarantee the coordination of the same people and material resources, and facilitate the cooperation of professionals in journalistic production (Salaverría and Negredo 2008). As collaboration involves these three aspects, they can be represented by the 3C Collaboration Model. *Communication*: the exchange of messages; *Coordination* or management of persons and resources, including technological resources; and *Cooperation* of operations in a shared space. Research on the 3C collaboration model, based on the principles of communication, coordination and cooperation, started more ten years ago in Brazil (Fuks et al. 2011). In 2008, Cook proposed the 4C Collaborative Model, with an added dimension, namely networking, which makes it possible for people collaborating to make *Connections*. This model was then enhanced by Costa et al. (2014), who integrated the role of social network software in contemporary newsroom production.

2.3   Ontologies of Contents, and the Semantic Web

From an early stage, research has explored a semantic orientation in journalistic production (e.g. the use of metalanguage, ontology and Semantic Web technologies), and how they can support the development of collaborative systems that support the functions of production within journalistic writing (Hollingsworth 1995). Indeed, even before attempts at the use of semantic technologies, newspapers had already made significant investments in their news management systems and undertaken considerable standardization efforts in order to facilitate interoperability (Troncy 2008).

One of the main standardization frameworks in the journalism domain is that of the International Press Telecommunications Council2 (IPTC), an international consortium of news agencies, editors and newspapers distributors (García et al. 2006). IPTC NewsML-2 supports semantic underpinning in its standards and is an issue of great importance and relevance to our work. In research studies of interest to us, Huovelin et al. (2013) analyze data collected from the Internet and to identify information that has a high probability of containing new information. The identified information is summarized in order to help understand the semantic contents of the data, and to assist the news editing process. García (2014) also presented how semantic technologies make it possible to go beyond Digital Rights Management and, because it is possible to model copyright through the whole media value chain, manage media rights from creation or remix to end-user consumption. In related work, Fricke and Thonsem (2014) identify the necessity to integrate the department-oriented search and editing of news of a TV-newsroom, into a more integrated process, bridging this traditional division. So they propose a news workflow modeling oriented by semantic annotated fragmented media.

More recently, Christensen and Jacobsen (2017) presented a semantic news aggregator system called News Hunter, a semantic news aggregator. The system extracts named entities and keywords from incoming text, and then stores and uses these to gather data from other resources capable of presenting journalists with up-to-date background information on incoming news messages, as well as live-updated information when writing new stories. Moreno-Schneider et al. (2017) present a prototypical content curation dashboard, to be used in the newsroom, and several of its underlying semantic content analysis components (such as named entity recognition, entity linking, summarization and temporal expression analysis). The idea is to enable journalists (a) to process incoming content (agency reports, twitter feeds, reports, blog posts, social media etc.) and (b) to create new articles more easily and more efficiently. In earlier, work. Palmonari et al. (2015) proposed a framework for news reading using data context to support readers interested in data-supported stories for data driven journalism and journalists in the newsroom. To support these functionalities, they applied semantic technologies in the news domain to connect articles based on the co-occurrences of named entities or to extract relations among entities, which they called relational data journalism.

3 Methodology

This study focuses on the evolution of newsrooms and their support systems for news production in contemporary journalistic writing, with a look at information virtualization, social networks, cloud computing, new mobile devices such as tablets and smartphones, and the environment (de Mendonça Jorge et al. 2016). In so doing, we sought to identify the requirements and functionality for constructing an interaction model and the construction of a framework model for the provision of information in newsrooms.

3.1 Newspaper Visits

This article is based on the results of field trips undertaken over sixteen months between 2015 and 2016 in the newsrooms of *Correio Braziliense* (Brasília/Brazil), *O Globo* (Rio de Janeiro/Brazil), *La Nación* (San Jose, Costa Rica, and the BBC and Reuters (London, UK). In each site, visits lasted around 30 h; researchers observed journalists and editors in action as they worked on stories, conducted editorial meetings and produced news (Cordeiro and Lessa 2014; Medeiros Neto et al. 2016). The profile of the organizations and their business structure are summarized in Table 1.

Researchers also conducted and recorded in-depth interviews with reporters, editors and support staff of the five large media companies. The visits had the objective to collect the functional and non-functional requirements of the proposed model. Formal conversations during the visits were complemented with informal meetings outside the newsrooms, giving greater insight into journalists' professional culture and identity (Undurraga 2017).

3.2 Comparison Between the Three Main Media Organizations Visited

Virtual collaborative work environments based on the 3C/4C model can be used in at least three different ways: one as a tool for Communication, as support for Cooperative work, or as a support for the model of journalistic production Coordination in Semi-faceted (or blended) format. Moreover, the implementation of another C, the C of Connection, is favored by ubiquitous communications.

Accordingly, one of the goals of the visits was to evaluate the workflows in the three visited newsrooms, as detailed in the previous section. Five aspects were observed: integration and convergence, channel distribution, semantic and workflow perspectives. Workflow aspects can be summarized in the following:

Table 1 Comparison between the three news organizations visited

	O Globo (Brazil)	La Nación (Costa Rica)	BBC (United Kingdom)
Newsroom workflow	Adopted an integrated and partially convergent editorial office, with the print newspaper editorial areas in charge of producing material for both print and digital versions	Has a fully integrated and convergent newsroom, including planning, generation of content, and publication	This newsroom in London, called The World's Newsroom, reflects a merger under the same roof of BBC's international journalism with state-of-the-art technology. The space was created to privilege multimedia production for multiple platforms: TV, radio, and online media, in various parts of the world
Integration/ convergence	Convergence processes allow work between more channels, media and integration processes. Convergence is happening according to four factors (4P): Processes; Platform; People; And Products	The first attempt at integration and relevant convergence was in 2007, and it was strongly resisted due to the culture of print journalism, to the detriment of IT use. But gradually the convergence and the integration were advancing and being absorbed by the professionals Workflow is divided into a main block, Integrated Drafting, and the second block, which deals with Documentation Gestation, including Documentation Center	The BBC has a 24/7 news broadcast, so a channel which is always broadcasting News, however, there is a big studio for many channels, all of our cameras are robot-controlled, remote controlled, so the camera movements are preprogramed into a document, an XML document actually, that describes a series of camera movements in relation to a script
Distribution channels	Printed, Website, Social Network and News Services	Printed, Website and News Services	TV, radio, Website, Social Network and News Services
Semantic perspective	– The processes of attribution of key words of the news were designed in Zunit; – Uses tools to define an ontology environment (Menthor, Protegé); – Use of ontologies in the G1.globo.com,	– *La Nación* was exploring ways to transform content; – Work was done to enable a semantic web searching on *La Nación*'s platform for content management	The BBC has been making use of semantic technologies in its internal content production systems since 2011. This enabled the publishing of news aggregation pages 'per athlete', 'per sport' and 'per event' for the 2012

(continued)

Table 1 (continued)

	O Globo (Brazil)	La Nación (Costa Rica)	BBC (United Kingdom)
	globoesporte.com, and ego.globo.com websites; – There was no well-planned planning for organization migration to Web 3.0		Olympics – something that would not have been possible with hand-curated content management. Semantic underpinning of content is being rolled out on BBC News from early 2013 to enrich the connections between BBC News stories, content assets, the wider BBC website, and the World Wide Web[1]
Newsroom workflow perspective	The development in four phases of the integration, to carry out the production of printed and digital news: (a) evaluation of possible news of the day; (b) construction of the contents identified as possible news; (c) evaluation of the news produced; And (d) formatting for printed newspapers and the provision of web pages	*La Nación*, envisages five phases of the convergence of its writing and installations. It surpassed the first phase of digitization, and then (2) deployed online structures, and went through Phase 3, the physical integration of the essay and its facilities. Phase 4, the development of new languages for the management of the production process, faces a process of comings and goings, with successes and failures, as highlighted by interviews during our site visit in 2016. In relation to Phase 5, that of the total fusion of structures, integration of means and production processes, it can be said that it is still in the planning stage and pending adaption of the previous phases	The advances identified at the BBC in London are related to the perception of mutations for a more complex and collaborative work environment. The newsroom is being triggered mainly by the introduction of technologies and digital transformations, use of tools and frameworks capable of integrating the Internet with routines of the news production process and use of Semantic

[1]More information available at: http://www.bbc.co.uk/blogs/internet/entries/63841314-c3c6-33d2-a7b8-f58ca040a65b

- *Globo*: The development in 4 stages of the integration between production of printed and digital news: (a) evaluation of possible news of the day; (b) construction of the contents identified as possible news; (c) evaluation of the news produced; and (d) formatting for printed newspapers and the provision of web pages.
- *La Nación*: the development of new languages for the management of the production process faces a process of comings and goings, with varying success, as highlighted by interviews during our site visit in 2016.
- BBC: The newsroom is affected mainly by the introduction of computational systems technologies and digital transformations, use of tools and frameworks capable of integrating the Internet with routines of the news production process and use of semantic technologies.

4 From Newsroom 1.0 to 4.0

Since the post-war era, the media business model has remained relatively unchanged around the world, concentrating on producing single media content by selling media products and advertising spaces in the media. The main feature has continuously been broadcast and the diffusion, but always having the newsroom as the heart of the newspaper, be it big or small. Newsroom integration progresses everywhere. In this integration process the effects on the daily routines of media professionals have received less attention. Exceptions from this general observation are studies on newsroom convergence that focus on the effects of converged production on journalists (von Rimscha et al. 2016).

That newsrooms need to attend the requirements of the next generation digital news industry is based on the assumption that consumers, now viewers, will become participants too. This fact implies the need of new features to support interactive devices, content adaptation, and management for new distribution channels (García et al. 2006). The need for new features for content and production management is important for the digital[1] news industry. In this article three newsroom features are proposed: Newsroom 1.0; Newsroom 2.0; and Newsroom 3.0, the aim of which is to improve news production and knowledge management (Medeiros Neto et al. 2019).

---

[1]To see more about workspace design for editorial area to modern media companies see the following link: https://www.americanpressinstitute.org/publications/reports/strategy-studies/matter-of-space/single-page/.

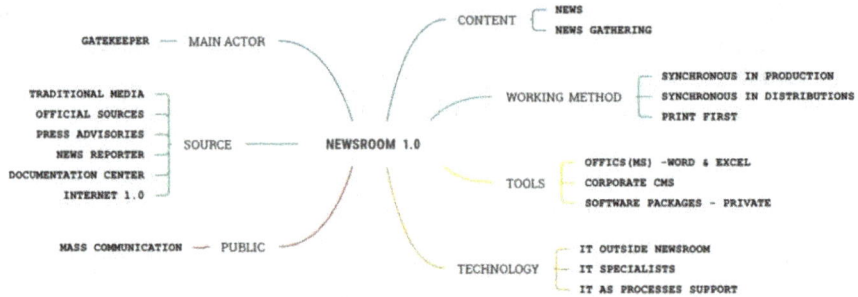

Fig. 1 Newsroom 1.0 features *Source* Authors (2018). Free translation

4.1 Newsroom 1.0

The design of this first newsroom feature was the result of observation of its culture, so it is vital for understanding not only the production of knowledge in the context of media convergence, but also the shifting balance between politics, economy and the media in the online era. The ethnography of *O Globo*, *La Nación* and BBC showed that traditional approaches to the political economy of the media (de Mendonça Jorge et al. 2016), tend to privilege the power of structural forces, e.g. media ownership and professional conventions (Undurraga 2017). Newsroom features are mainly composed of main actors, information sources, news audience, the content, the working methods, the tools used in the newsroom and the way technology is used in the newsroom. All these features are built to produce news that is distributed at broadcast channels. Figure 1 shows these newsroom features, all encapsulated in the umbrella term Newsroom 1.0.

4.2 Newsroom 2.0

This second newsroom features addresses the work of journalists in newsrooms producing content for multiple media: print, radio, television, the Internet and others. Specifically, the features show the change in journalistic practice and workflow in the newsrooms of media companies and business groups.

The visits to the cited newsrooms offered a useful script through which to observe the power dynamics within newsrooms and the way historical legacies shape news organizations, but visits also paid attention to the role of technologies, for examples tools and information systems in use by the various journalistic agencies in news production, and how they are impacting the people, material resources and business.

Virtual collaborative newsroom work environments, based on the 3C model can be used in at least three different ways, one as a tool for communication, next as a support for cooperative work, and lastly as a support for the model of journalistic production coordination in Semi-faceted (or blended) format.

The main differences in respect of Newsroom 1.0 are given by the fact that a new actor (the Web Editor) appears, whilst Wikis, Databases and Social Networks are added to the information sources, and the community is added to the public features category. In Newsroom 2.0 the content is scrutinized before being published and the working methods are synchronous processes to produce mainly printed news. Moreover, the tools used to produce news are software packages whereas the IT used to produce the news is a distinct department outside of the newsrooms. All these features are built in order to produce news that is distributed through broadcast channels.

Journalistic companies that fully adopt the newsroom structure of Newsroom 2.0 adapt their news production processes through online platforms where journalists now share a much greater pool of information. This has happened largely through technological innovation by introducing a second desktop screen: most journalists now sit facing two computers, one for writing up their reports, and another which broadcast real time news of others information of the company. From this perspective, attributes such as working in teams, collaborating and transmitting knowledge are now seen as positive differentials to employees. To assist in this process, there are electronic collaborative systems that assist employees "in the different phases of social interaction within the teams: communication, coordination, cooperation/ collaboration and connection" (Schauer and Zeiller 2011). Figure 2 shows these newsroom features, all encapsulated in the umbrella term Newsroom 2.0.

Furthermore, in Newsroom 2.0 digital media open up a space of experiments and actions for communication. Artificial intelligence (AI) embodied in automated journalism manifests itself through processes of data access, interpretation, writing and distribution, and is already a fact in the major newspapers and news agencies of the world (Squirra 2016). The visited news organizations, *La Nación* and the BBC

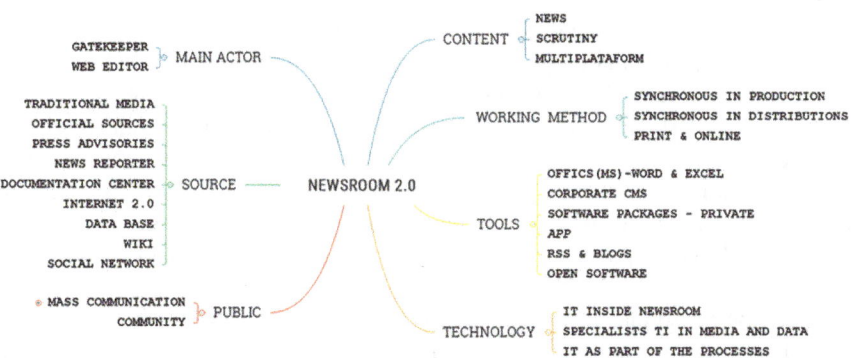

Fig. 2 Newsroom 2.0 features. *Source* Authors (2018). Free translation

could fit the Newsroom 2.0. However, projects in development in these organizations allow us to say that they are migrating to a Newsroom 3.0 model, which will be described next.

4.3   Newsroom 3.0

A new leap forward, a new structure in terms of complexity, begins to happen as new news-producing agents present themselves via blogs and personal pages—all accompanied by a greater connection of people to the social networks using smartphones, a radical increase in the number of interconnected sensors, all whom produce and seek information, everywhere and at any time. Newsroom 3.0 features are thus established by "the digital transformation of the interconnection of people and networks that exchange information in two-way flow whenever possible" (dos Santos 2016). Technologies have made it possible to increasingly facilitate editing, curating from various sources, and developing, through mash up practices, other products that were not originally from a single platform. This creates not only a job market, but also a very interesting digital experience.

The third newsroom feature shows how real-time technology is dramatically changing journalism. Due to the speed of information, one does not have time to build a report in days or weeks, as might have been previously the case. However, at the same time, journalism guided by data, with the possibility of abundant data and information, allows us to practice more accurate journalism. On the other hand, production of news is only informative and not reflexive. These perspectives force us to use a collaborative environment to accelerate news production in newsrooms.

A new mode of news production thus emerges that incorporates new agents, with an increase of information flows generated precisely by these agents and IoT devices, all of which are now news prosumers (dos Santos 2016). These changes create openings for new ways to collect, create and disseminate news as well as to interact with news audiences in different ways (Deuze 2006).

A new step in terms of complexity, represented in part by Newsroom 3.0, begins to happen as, in addition to the original inhabitants and newcomers, the prosumers of news content appear on the scene. Accordingly, Fig. 3 shows the features of Newsroom 3.0. The main characteristics that reflect its evolution from Newsroom 2.0 are: the primacy of digital over print media, the presence of asynchronous processes, an increased curatorship (content created by the audience), as well as an increasing tendency to use social networks and IoT devices as news sources, all coupled with an incipient but growing use of augmented reality to convey news content.

After the consolidation of social network software in Newsroom 3.0, the connection has become an essential function, by "*allowing people to make connections with the content and among others*" (Schauer and Zeiller 2011). This context considers that collaboration occurs when people, who have at their disposal great autonomy and responsibility with the collective, work together sharing goals and

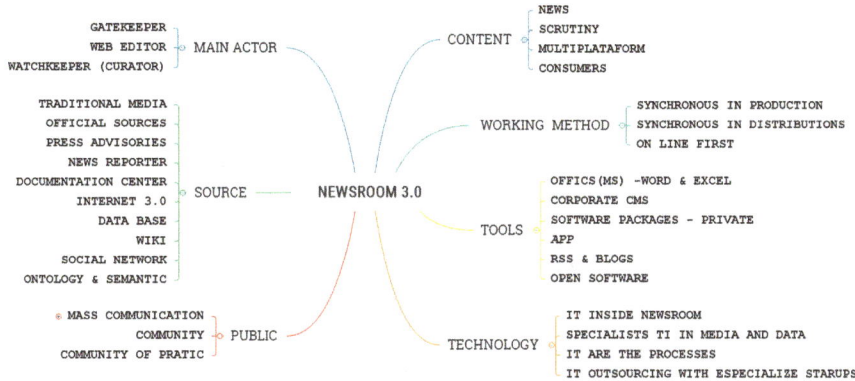

Fig. 3 Newsroom 3.0 features. *Source* Authors (2018). Free translation

commitments while motivating themselves intrinsically (Schauer and Zeiller 2011). Figure 3 shows these newsroom features, all encapsulated in the umbrella term Newsroom 3.0.

4.4 Newsroom 4.0

The next step, the Newsroom 4.0, is to become totally digital and virtual. A platform digital, Newsroom 4.0 is an infrastructure with affordances offering diverse kinds of information, communication, and other services. Digital Journalism is more than a platform since it goes beyond creating opportunities to produce, publish, and engage with content. A commutation company that operates with Digital Journalism do not produce and publish content themselves, and thus do not define themselves as publishers. Instead, they operate with a new business model in which they provide a Newsroom 4.0 on which individuals, communities, and institutions can communicate, connect, collaborate for getting, and publishing information (Ekström and Westlund 2019).

There is no more space for printed media. Even totally digital and virtual small media companies that are competent will survive. In an environment prone to epidemics and pandemics the personal physical isolation is something that will become increasingly frequent, and this will change our lives forever. In this scenario the online commerce will be the first option. Education will be online for all life. The tourism market will change to a more sustainable and safe form, without fuel consuming long trips or overcrowded tourist attractions. Tourism will be an immersive virtual reality experience, with virtual souvenirs sold in the Newsroom 4.0. Business trips will be replaced by videoconferences. The health system will be distributed; telemedicine and home care will be reality. The supply chain will be virtual, substituted by digital instructions to print 3D parts and components.

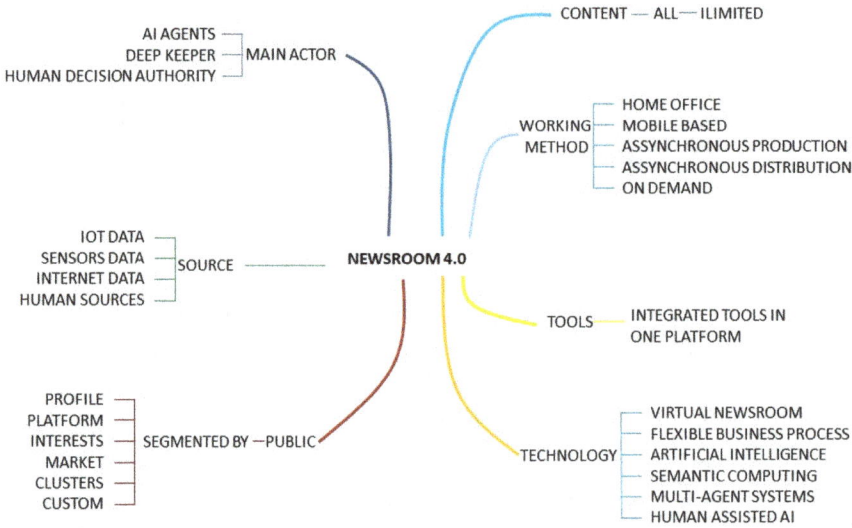

Fig. 4 Newsroom 4.0 mental map (in progress). *Source* Authors (2018). Free translation

Our real needs to sustain our lives will become simpler, the basic to survive, but our virtual needs for our virtual life in our virtual environment will be increasingly sophisticated and environmentally friendly. Besides our physical isolation, we will be more virtually connected and will have more time to spend our lives with whom we really care. In this future scenario, that is becoming current with the Covid-19 pandemic, Newsroom 4.0 will change dramatically. Figure 4 shows these newsroom features, all encapsulated in the umbrella term Newsroom 4.0.

In this new world, with increasing spatial distribution among persons, increasing connectivity and automation, the cooperation, coordination, communication and connection will be the keys that will leverage virtual works environment. But in this virtual environment, it's fundamental that the principal actor continue to be the human-beings to decide the actions in the real and virtual worlds.

We must be in control of everything. To keep this goal all automated process must obey human guidelines established by compliance, governance and risk management.

The main features that characterize Newsroom 4.0 are that the main actor must be a human being, helped by AI agents and deep keeper that will have algorithms to reinforce the human generated guidelines to the editorial line, the ethics values to be respected and the great quality on all the work and the results made in the Newsroom 4.0. In this context, the glue that connect distributed home offices, some of which will be mobile, but all working on demand in an asynchronous and distributed fashion is the capacity to account the business process, to achieve the compliance and governance goals of all involved in this system, but without missing the risk management, which involves the risk management of time (opportunity), costs and news quality.

In this work environment the tools must be integrated in one platform to better manage the newsroom business process, which will use many technologies like virtual newsroom, flexible business process, artificial intelligence, semantic computing, multi-agent systems and human assisted AI to mention some technologies. All these technologies will help to process a myriad of data and information/ knowledge sources beginning with the most important the human source of knowledge and information complemented by IoT, Sensors and internet data to allow the production of any type of content, chosen by humans, for a public segmented by profiles, platform usage, interests, market, clusters or custom.

In order to enable an environment with the aforementioned features, we will model in the next section the fundamental of the Newsroom 4.0 framework.

5 The 4C Model

The 3C Collaboration Model, whilst having many uses and applications, has limitations when it comes to dealing with the case of multiplatform environments, or in that of projects employing social networks. In the particular case of journalism, the 3C Collaboration Model hasn't fully explained the environment of a newsroom. To address this issue, in this current section we present an enhancement to the 3C Model, namely the 4C Collaborative Model.

5.1 The 3C Model

Although the 3C model was elaborated in order to model collaboration in a Computer Supported Cooperative Work (CSCW) context (Fuks et al. 2011) it exists before the ICT advent, as seen in Fig. 5.

In a local/centralized system, the 3C model has three main dimensions (Fig. 6):

1. *Communication*

   a. Develop learning, understanding and consensus about something
   b. Create content to spread though some channel
   c. Serves as a manner to share ideas to achieve cooperation and makes the coordination possible.

2. *Cooperation*

   a. Create content collectively
   b. Uses communication to minor conflicts, achieve consensus
   c. If we want to cooperate, we may use methods and best practices to achieve better results.

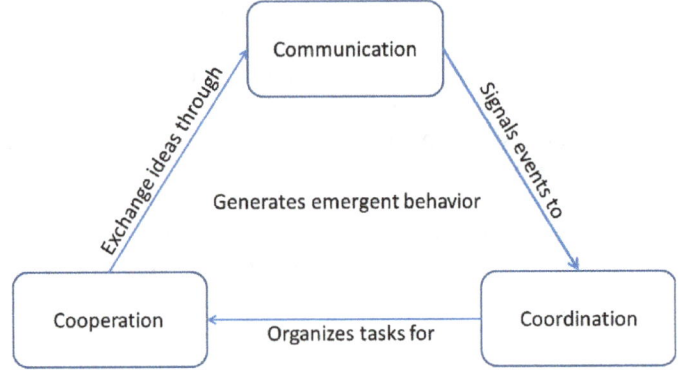

Fig. 5 3C model before the advent of ICT. *Source* Elaborated on the basis of Fuks et al. (2011)

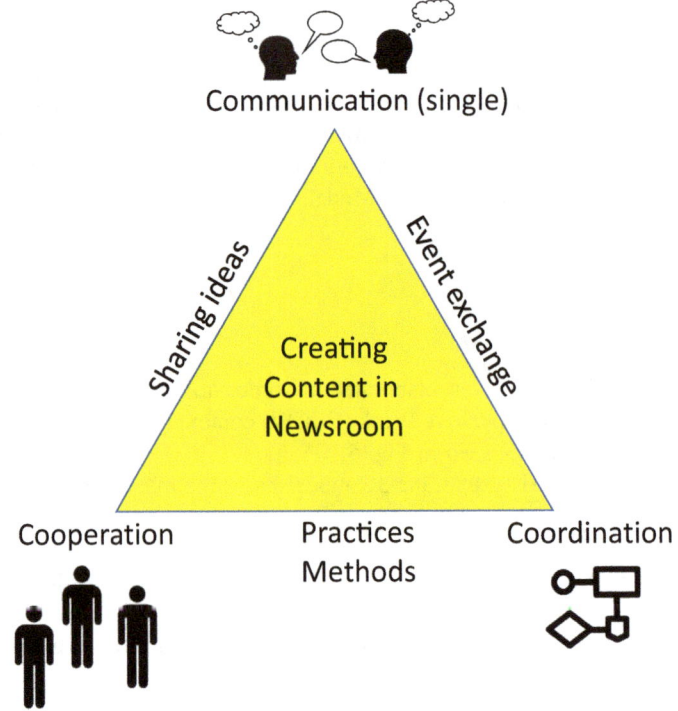

Fig. 6 3C model as a local/centralized newsroom. *Source* Authors (2018). Free translation

3. *Coordination*

    a. Plan to achieve goals

    b. Receives events produced by cooperation through communication to synchronize cooperation.

However, when ICT is introduced, communication may be split in 2 parts: data/Information and knowledge. With the increase of data/information in the Internet, one fundamental requirement to handle this requirement is the capacity of computer software to understand what is being done, what is being processed, stored, retrieved and transmitted. Fortunately, semantic web technologies give us the basic tools to handle these requirements.

5.2 The Revisited 4C Model

Although the 3C model worked with the communication channels provided by the Internet (e-mail, www, etc.), it was enormously facilitated by the proliferation of social media. Each social media has its own way of communication, with its own language and its own public. If there is a need to reach a large audience and use different ways of communicating, one needs to use multiple media channels. In this way, a new actor appears in the 3C model, the connection, which generates our interpretation of the 4C model, shown in Fig. 7. We opt to represent our model as a three-dimensional tetrahedron because visual representations are better way to understand the relations of more than three entities (Fuks et al. 2011).

Figure 8 shows tetrahedron framework planning, in which each face of the framework corresponds to one dimension of the final newsroom.

The dimensions of the framework are:

- The newsroom itself;
- The knowledge;
- The workflow; and
- The social media.

The four dimensions of the framework correspond to a view from the perspective of an ICT Professional with four management systems dimension—WMS (Workflow Management System) to support Workflow and information flow, CMS (Content Management System) for the elaboration of the content, SMMS (Social Media Management System) for the management of media channels to collect content, interact and distribute the news; and KMS (Knowledge Management System) to give semantic support, memory, knowledge and learning (da Fonseca et al. 2018; de Deus et al. 2018). These four dimensions are detailed in the next subsections.

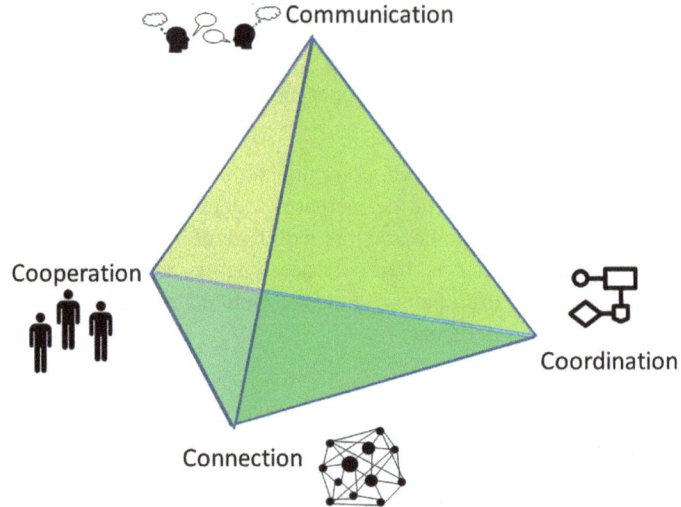

Fig. 7 The tetrahedron, a revisited 4C model *Source* Authors (2018). Free translation

Fig. 8 Tetrahedron framework planning. *Source* Authors (2018). Free translation

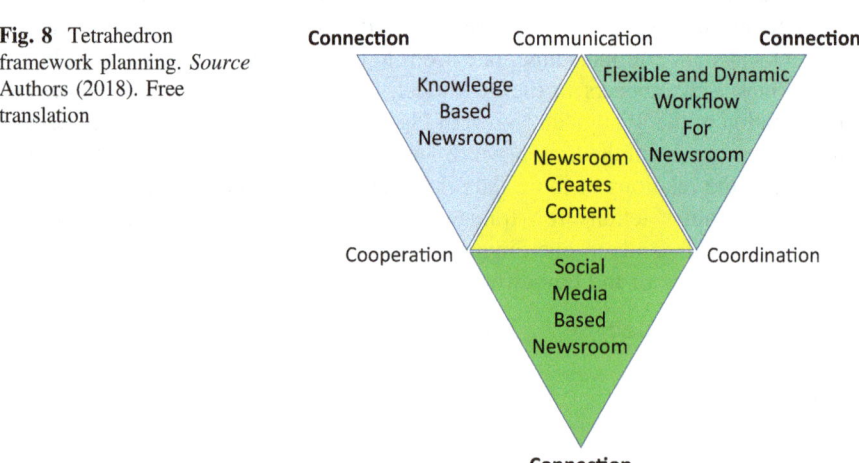

5.3   The Business Domain Dimension

The tetrahedron framework is applicable to any business domain. If one thinks about a conventional Information System (IS) to a specific domain, the face of the business domain is sufficient. Some domains require more or less *communication*. If the IS is at the operational level, *communication* is a minor dimension, however, if it is on the knowledge level *communication* is a key element (Laudon and Laudon 2015).

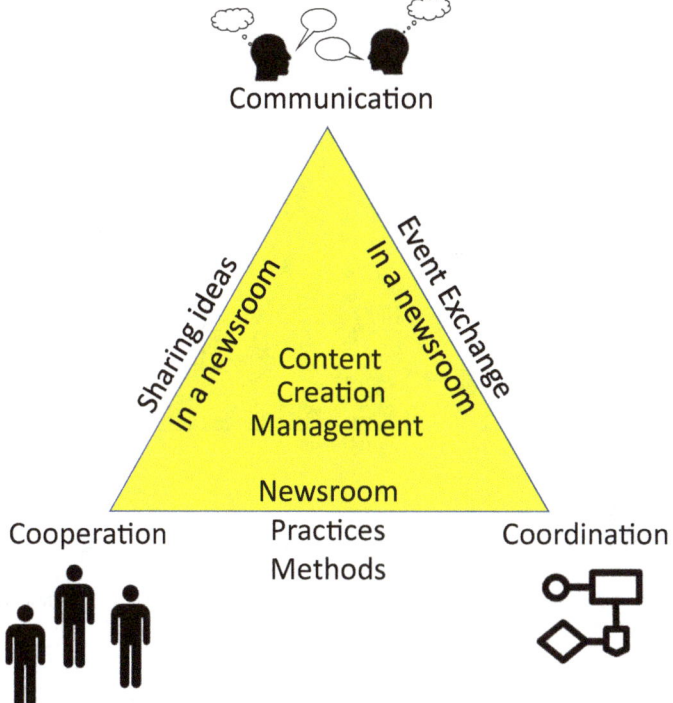

Fig. 9 The tetrahedron framework face of the business domain dimension. *Source* Authors (2018). Free translation

So, to explore the framework on all of its dimensions, an enterprise knowledge level IS will be used. In this case, the people who work in this level need to *cooperate* sharing ideas of the enterprise business domain to achieve its goals. Also, they have to *communicate* in order to achieve consensus and diminishing conflicts. To better achieve its goals, this team may use methods and best practices to *coordinate* their activities. Figure 9 shows the business domain dimension.

The following three dimensions, showed in the next three subsections, emerged with the easy and infectious connectivity offered by social media.

5.4   The Workflow Dimension

The workflow dimension represents the enterprise activities organized as a process with its activities, events, gateways etc. to achieve some goals. The face of the workflow dimension has common edges with the others faces of the tetrahedron framework. Figure 10 shows these edges.

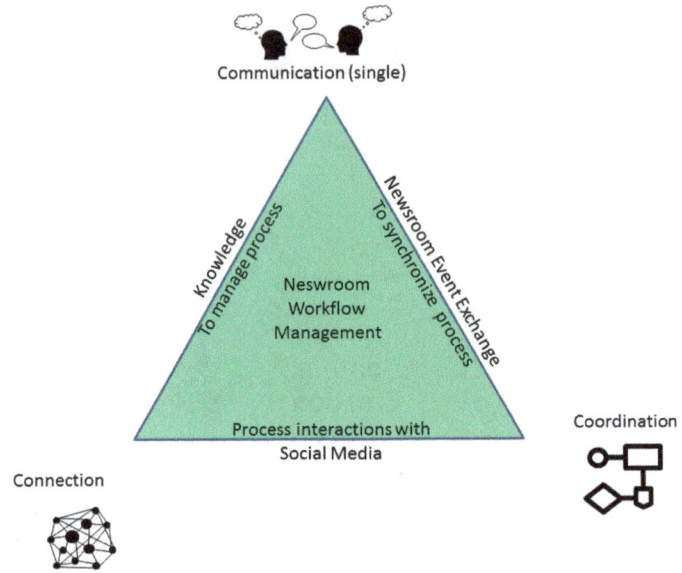

Fig. 10 The tetrahedron framework face of the workflow dimension. *Source* Authors (2018). Free translation

One edge is shared with the Business Domain Dimension. Traditional Business IS implements the process flow coding it in the IS computer program. However, changing the code to adapt or create another process in IS takes time and resources of the enterprise. So, a conventional SI freezes the process for some amount of time that prevents the company from adapting rapidly to market changes, decreasing the organization´s flexibility. So, this dimension implements process in a dynamic way, in such a manner that process can be created, updated, modified, adapted or deleted on the fly.

Dynamic and flexible process must comply with strategic business process that enforce high level standards to keep the compliance, ethics and governance of risk management of the media enterprise accordingly with its own editorial line at the operational level to produce high quality content with an accountable operational flexible business process.

As it has an edge in common with the knowledge dimension, the process would be mapped to an ontology, like the BPMN (Business Process Management Notation) ontology, and all of its elements and flow could be annotated. This annotation represents the knowledge behind the process and the logic that governs it. This knowledge may be used to help the manager to adapt or to build new process.

The other edge represents the challenge of this face and framework. The question that one poses in this context, is how does social media interact with the IS process in order to achieve its goals? The immediate application is the synchronization of the interaction of all those involved in the IS by means of events and conclusion of the tasks of the business processes. A most ambition challenge is that this process could be adapted according to users' interactions. In this case, the knowledge dimension helps this facility using the knowledge stored on it using its inference mechanisms and even AI techniques.

5.5   The Knowledge Dimension

This face represents the intelligent face of the tetrahedron. It stores the knowledge as domain ontologies and the semantic annotations about the others framework dimension. It also has an inference engine that implements algorithms and other AI techniques to help the users and managers of the other dimensions. The common edge with the workflow dimension illustrates that it helps this dimension storing a business process ontology like the BPMN ontology (Annane et al. 2019), as well as process-specific semantic annotations. The shared border with the business domain dimension stores the domain ontology of the business and all the semantic annotations about its data/information and the use of it. Finally, the frontier with the Social Media dimension stores the knowledge about the social media channels, the languages it uses to communicate, the audience that it reaches, in order to better communicate and to reach and to interact with the right audience at the right time. Figure 11 shows this face.

5.6   The Social Media Dimension

The social media dimension is the novelty of this framework. It integrates the business social media used by the IS in an unique system that can manage this media channels. This dimension is useful to be independent of social media monopolies, like Facebook, Google or the Chinese Tencent. One initiative in this respect is that of Facebook, which is announcing its intentions to integrate WhatsApp, Instagram and Messenger to allow people to communicate across the platforms (Isaac 2019).

Because this dimension has common borders with the business domains dimension (Fig. 12), the framework dimension and the knowledge dimensions it has the sufficient and necessary tools to integrate media channels, facilitating their joint management.

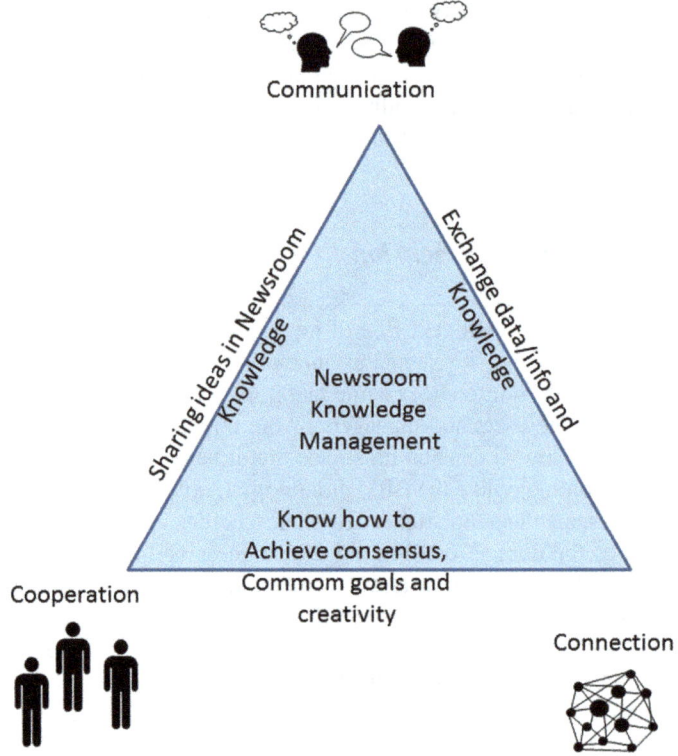

Fig. 11 The tetrahedron framework face of the knowledge dimension. *Source* Authors (2018). Free translation

6 The Framework Architecture

The four vertices represent the revisited 4C model, the Communication, the Cooperation, the Coordination and the Connection Dimension. The tetrahedron's faces represent the systems that integrates the Business Domain, the Workflow, the Knowledge and Social Media Systems that are orthogonal to the Vertex Dimensions. Moreover, the edges represent the interactions among the faces to achieve the vertices' goals.

Figure 13 shows the resulting tetrahedron framework applied to a newsroom management system. To implement the framework the Django Python Framework was extended to support KMS, WMS and SMMS. We extend the Interactive Knowledge Stack (IKS) (Behrendt 2012). The IKS is an open source community project focused on building an open and flexible platform for a semantically enhanced CMS. Originally IKS had two main stacks, the conventional CMS stack

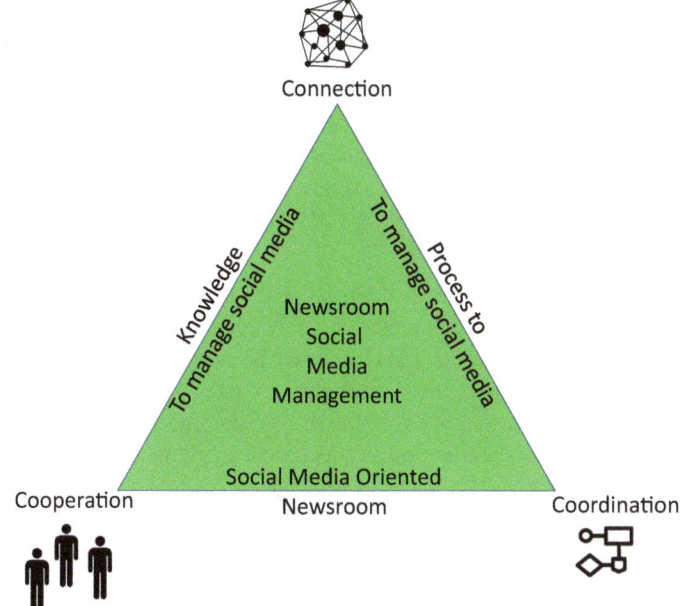

Fig. 12 The tetrahedron framework face of the social media dimension. *Source* Authors (2018). Free translation

Fig. 13 The tetrahedron framework in a nutshell. *Source* Authors (2018). Free translation

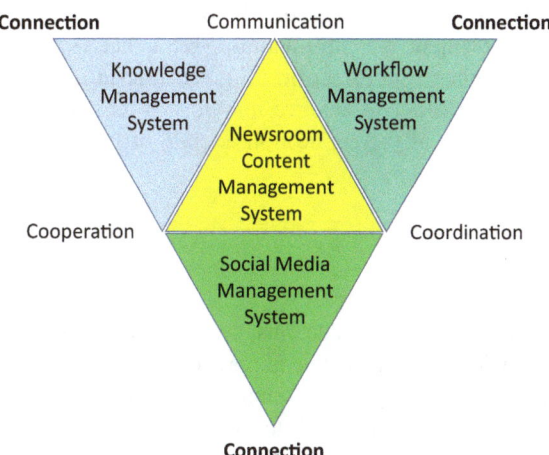

and the Semantic stack (KMS), one stack communicating with each other and a semantic interface. We extended IKS with another stack, the Business Process Management stack, with Newsroom 3.0. Now, we add another stack, the Social Media Management System (Newsroom 4).

6.1 CMS

The CMS was meant to facilitate the editorial process of news creation. This system provides some standards that integrate the analysis of XML files. Our framework uses the NewsML standard from IPTC to describe news information content so it can be widely distributed and reused on Web sites and other media. It is also worth mentioning that NewsML standards are based on a common framework—the News Architecture—that is independent of any technical implementation (IPTC 2017).

6.2 WMS

Newsroom 3.0 implements the Workflow dimension as a Workflow Management Systems (WMS) that assists the modeling of news business processes in a flexible newsroom, in which process could be created or modified on the fly. Thus, we created a light domain ontology for this newsroom by using the methodology 101 and Protégé modeling tool. The BPMN standard (Annane et al. 2019) was also used to implement the information system that would automatically suggest the most appropriate role to perform a given task in a news production process of a newsroom. In order to achieve this goal, a tool was developed to aid in the semantic annotation of the application domain of the processes (da Fonseca et al. 2018).

6.3 KMS

Newsroom 3.0 implements the Knowledge Management System (KMS) is the knowledge stack of the framework. It uses a Relational DBMS to store n-quads tuples of ontologies and semantic annotations in RDF and OWL format. It also uses Python 3.x and RDFLib to implement all the elements of the knowledge stack.

6.4 SMMS

Newsroom 4.0 implements the Social Media Management System combines the CMS, WMS and KMS to manage Social Media Channels in order to receive inputs, process them and create outputs to Social Media Channels using the appropriate language for each social media channel accordingly with its audience. To achieve this goal the SMMS stores in the KMS the ontology of the social media channel. This ontology could have the main audience (for instance, the audience age or business), characteristics of the language (Twitter limits its messages to 280 characters), and so on. At the WMS the journalists could make the appropriate process to generate a tweet automatically handling the activities, which consist mainly of discovering the content, filtering and contextualization, verification of truth and publishing to media channels. Of course, each process is adapted to the news subject, sources, audience and so on.

6.5   News as a New Social Media

At present, there is already a great amount of journalistic material that is being produced, either within newsrooms, agencies or by readers and content generators. Another fact is that it has become impossible to follow developments more accurately without the support of ICT. Moreover, new questions arise: how does one use the very interfaces of social networks? How does one monitor and check all content contributors, and still match the interests of this new reader? Certainly, a journalist unprepared to handle data-driven journalism (data journalism) or even the lack of a data analytics professional will be at a disadvantage. Much akin to a manager who does not know how to deal with the flow of information from centralized or distributed newsroom, based on Information System or Social Media Management System—SMMS, it will bring limitations to the production process of journalistic materials.

Journalism has almost always sought to exploit the capabilities of new IT platforms in many countries, even before social networking becoming in a very short time an indispensable tool of professional value, used daily to monitor events, find sources and verify information. The Brazilian elections of 2018 were a breakdown of paradigms, in the dispute between these and the mainstream media. However, journalists and media professionals continue to face a number of challenges in using collaborative social networks and information systems. Fundamentally, the problem is one of scale and that of incorporating the use of emerging technologies such as data scraping and database analysis, workflow and Semantic Web, which have changed the journalistic landscape in a very short time (Thurman et al. 2019).

The point is: it is almost impossible to handle such a myriad of social media channels manually; this process has to be automated. Beyond that, journalism has to resume its primacy by not only learning to deal with social networks, but by enhancing their use through creating new social media news. Another alternative is the integration of social networks such as Facebook, Instagram, Twitter, YouTube, etc.

6.6   Application Beyond 3C Model to 4C Model

At this point the amount of information available to journalists is considerable and they have to handle all this in a very short time. Therefore, this new newsroom needs to incorporate tools exploiting Semantic Web technologies into the world of professional journalism, in order to improve quality and productivity in the newsroom. Table 2 synthesizes and summarizes the evolution from Newsroom 1.0 to Newsroom 4.0, with the application of the 4C collaborative model, in terms of use of the communications, coordination, connections and business model in the newsrooms.

Newsroom convergence is a multidimensional process that, facilitated by the widespread deployment of IT and telecommunications, affects the technological, business, professional and communicative spheres of the news production means, providing an integration of tools, spaces, working methods and languages previously disaggregated, so that the journalists elaborate contents that are distributed through multiple platforms, through the languages proper to each one (Salaverría et al. 2010). At the *communication* dimension the increasing ubiquity of the Internet and the multiplicity of mobile communications applications and their use in the work environment allows the evolution of the low communication from the old newsrooms to the high common of the contemporary newsrooms. In a similar way, the *connection* dimension is increased by the virtual face-to-face communications applications (social media). At the *cooperation* dimension the traditional newsrooms follow the industrial revolution model to press the news (paper), which imposes rigid restrictions (viz. a high division of labor and high synchronization to produce the news), but in the transition (2.0) traditional press systems co-exist with the online press, creating challenges with their mutual coexistence. The management (*coordination*) differentiating between the two types of newsrooms leads to a drastic reduction of the print media and opens up novel possibilities of online cooperation and coordination in the news domain, i.e. Newsroom 3.0.

Analyzing the evolution from Newsroom 1.0 to Newsroom 4.0, we conclude that the basic elements that remain in all 4 newsroom features and are the core of all newsrooms are: human beings (knowledge); the process to produce news (workflow) and the elaboration of the news (content). These basic elements were observed in the visits to the five media organizations and confirmed in the literature review. The managers in these news organizations always identified how good management of these elements led to satisfactory results.

To help manage the work in the newsroom we mapped 3 newsroom facets: the cognitive effort required; the news production, and the news itself. Figure 14 shows how these 3 facets can contribute to build a system that can handle the needs from Newsroom 1.0 to Newsroom 3.0.

The main activities that represent **human beings** are their capacity to understand the meanings of the data and information, the behaviors behind them and the context where they are inserted. Moreover, the cognitive facet of a newsroom can also be improved, thus expanding human cognition through the use of semantic

Table 2 From Newsroom 1.0 to 4.0

Dimensions of the collaborative model	Newsroom 1.0	Newsroom 2.0	Newsroom 3.0	Newsroom 4.0
Communication	Low within and low outside of the Newsroom	Medium inside of the Newsroom and Low outside of the Newsroom	High within and medium outside of the Newsroom	High within and outside of the virtual Newsroom Intensive use of videoconferencing
Coordination	Synchronous of contents (average)	Synchronous (average) and Asynchronous (low) of contents	Asynchrony of contents (medium)	Asynchrony of contents (high). Commitment with editorial line, compliance, high level standards and risk management enforced by strategic business process mapped into operational business process
Cooperation	Industrial production (average)	Sharing production (high)	Collaborative production (low/loss of authorship);	Collaborative production (low/loss of authorship) with agent-based gatekeeper
Connection	Very low	Low inside and outside	Medium inside and outside of the newsrooms distributed	High inside and outside of the virtual newsrooms
Business Model and Profile	By access by users and advertising space of the Journal Economic: Group of companies; Multinational Corporations; Local and Global Communities	By access by users and advertising space of the Journal and hit on site Economic: Individual and Group of companies; Local. Regional and National	By online advertising and views Economic: Few Group and Multinational Corporations; Individual, Local and Global	Embedded advertising and views. Danger of poisonous fake news. Custom news Economic: Small and distributed

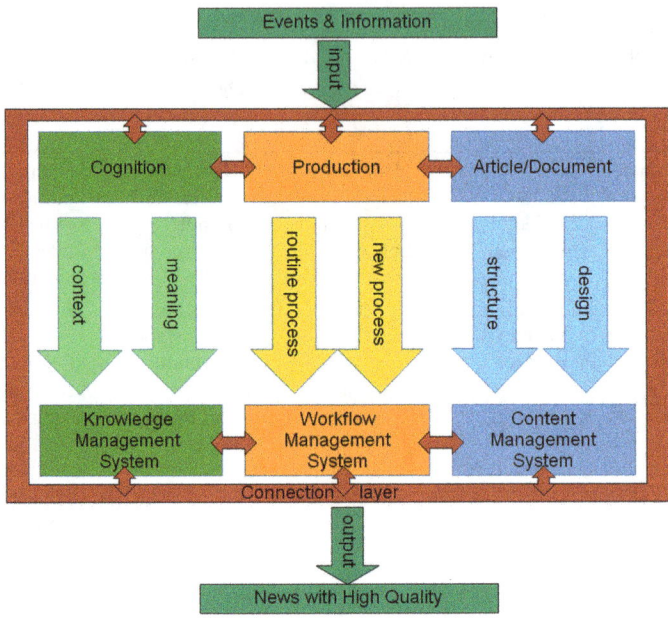

Fig. 14 Mapping the requirements to the newsroom needs. *Source* Authors (2018). Free translation

tools to extend their capacity to process enormous quantity of information using the meaning and the context of this information. This facet of the newsroom is handled by the KMS. In this second decade of the twenty-first century, journalists' work techniques and procedures, especially reporters and editors, have diversified, and this allows for more exact texts to be produced, depending on the favorable working conditions. The increased and ease of use of production tools, together with the Internet as well as the universalization of mobile devices are also democratizing the space of news content construction that could be offered by a KMS (Laje 2012; Medeiros Neto et al. 2016).

The production **process** in a newsroom reflects the strategy and goals of the team to produce relevant and timely quality news. It also reflects the technology that the organization can acquire and maintain which is dependent upon its resources. Newsroom managers believe that the consolidated workflow will facilitate the management of new projects and the optimization of the routines, giving support to the use of digital convergent media, according to a humanistic and social conception. So, the production facet of the newsroom is handled by the WMS that controls the daily process and can also make new process on the fly, according to the newsroom needs, given it the necessary flexibility.

The elaboration of an article (news/*content*), the composition of a page or of a communication vehicle are all composed of a work to design and structured it. In the visit to the BBC in London, it was confirmed by our interviewees that content

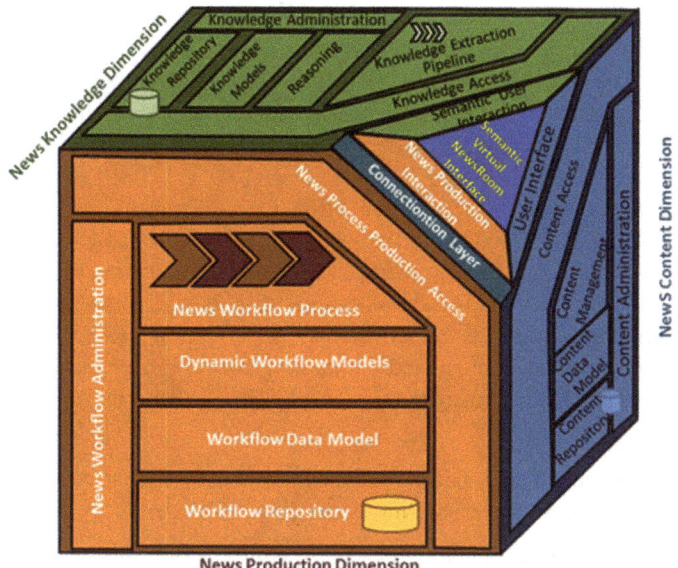

Fig. 15 Newsroom 3.0 Framework Model. *Source* Authors (2018). Free translation

productive by social networks may be introduced in their CMS. If, after appropriate vetting and verification, the news is confirmed the CMS then manages the BBC articles through the various phases in the newsroom, which includes the structuring and the design of the document itself.

Moreover, the people that work in newsroom nowadays communicate through many channels, from e-mails to social networks, Skype, WhatsApp, and so on. Thus, it is also necessary to integrate these communications tools into the newsroom in order to achieve a fourth dimension of the 4C model, the communicative dimension. This requirement is encapsulated by the communication layer, which integrates all the facets of the newsroom. The requirements could only map to a framework called Newsroom 3.0, but we go beyond this, stating that the Social Media dimension transcends the usual management of social media channels to be a new social media. This is why we called our framework Newsroom 4.0. Figure 15 presents our proposed Newsroom Framework Model, based on the IKS. It is structured along 3 dimensions):

- The News Content Dimension—Under this dimension, the framework is able to handle the elaboration of articles and documents. Moreover, journalists can also semantically annotate their work; retrieve works semantically related to their own articles; and put links to semantically related articles that may interest the reader (Oliveira 2004; Baños-Moreno et al. 2015; Oliveira et al. 2016). The news content dimension is implemented by a CMS (de Deus et al. 2018).
- The News Production Process (workflow) Dimension—this follows the Business Process Management Notation standards (Dailey et al. 2005) and was

modeled based on requirements gathered at the 3 newsrooms visited: *O Globo*, *La Nación* and the BBC. This dimension was designed to be flexible and to configure processes dynamically according to evolving newsroom scenarios and to also support content production in a potentially chaotic context (Ternai et al. 2016). A Workflow Management System (WMS) implements the news production dimension.

- The News Knowledge Dimension—this follows semantic standards of the World Wide Web Consortium (W3C) like the RDF and OWL standards (Antoniou et al. 2012). This dimension handles the annotation made in the news content dimension and news workflow dimension, persists that semantic information on a database and retrieves them through SPARQL queries, helping journalists to write content and build new newsroom process, respectively. A Knowledge Management System (KMS) implements the news knowledge dimension.

Apart from the three dimensions described above, the proposed model also comprises a connection layer and interfaces (Fig. 15):

- The Connection Layer is an interface that connects the newsroom environment to the cyberspace allowing external communications to other applications such as WhatsApp, e-mail, Twitter, Instagram, Facebook, telegram and so on. These external communications are new channels of communication to publish customized content. They make news available to different digital media like printed newspapers, online news, mobile news, etc.
- The Semantic Virtual Newsroom Interface is an interface to journalists that integrates the three dimensions of the framework. At this user interface, the journalists access the CMS, the WMS and the KMS. At the WMS they can model the process to produce high quality and reliable news. To help journalists in the modeling process the KMS offered semantic annotations of available processes and activities. Finally, the CMS follows the process build at the WMS. So, every journalist involved in this news production has a task and it appears to him/her with a deadline. At the end of the process, the news is available in the appropriate channel with opportunity.
- The workflow dimension interface, the knowledge management interface and the content management interface permit the communication among then and the Semantic Virtual Newsroom Interface.

7   Framework Model for Virtual Newsroom 4.0

To develop our News Production Framework based on the evolution of Newsroom 1.0 to Newsroom 4.0, we extend Newsroom 3.0, which is inspired in the Interactive Knowledge Stack (IKS) (Behrendt 2012). The IKS is an open source community, whose projects are focused on building an open and flexible technology platform

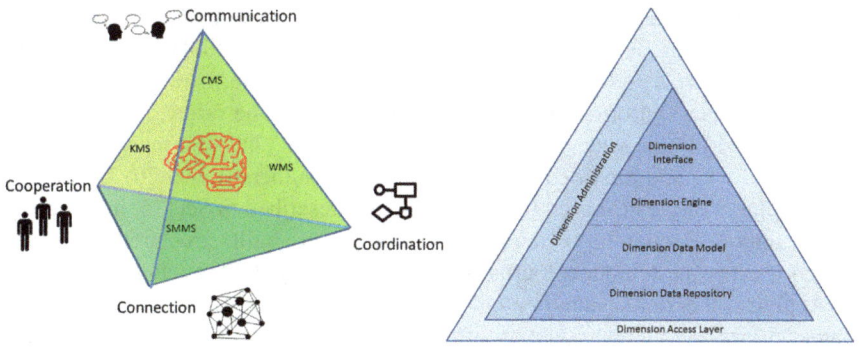

Fig. 16 Cyber journalism framework model. *Source* Authors (2018). Free translation

for a semantically enhanced CMS. With the resulting framework, one can design a virtual newsroom where the news flows among the newsroom stakeholders (chief journalists, journalists, press officer, news agencies, independent journalists, citizen media, etc.) in a custom-made process that adapts to the newsroom environment.

To help manage the work in the newsroom we mapped 4 newsroom facets: the cognitive effort required; the news production business process, the news itself and the new connection feature. Figure 16 shows how these four facets can contribute to build a system that can handle the needs for Newsroom 4.0, presents our proposed Newsroom Framework Model, based on the IKS. The framework shares the News Content, News Production Process, and the News Knowledge dimensions with the Newsroom 3.0 framework. However, it has an additional dimension and a kernel:

- The News Social Media Dimension, which connects the newsroom environment to cyberspace, allowing external communications to other applications such as WhatsApp, e-mail, Twitter, Facebook, telegram and so on. These external communications are new channels of communication to publish customized content. They make the news available to different digital media like printed newspapers, online news, mobile news, etc. (SMMS).
- The Kernel, the brain in the center of the tetrahedron, incorporates the framework intelligence, along with AI tools such as Intelligent Agent Management System and the inference engine.

8 Use of Newsroom Models in Scenarios

Prospecting newsrooms in scenarios. Whilst printed news is in a process of a rapid obsolescence and profit margins of news organizations are narrowing as a consequence, traditional news readership is also changing and even more so as youngsters are exposed to new ways of accessing information. The fact is that the news sector experiences a crisis never seen before, and multiplatform journalism seeks new business and production models (Becker and Guimarães 2016).

However, news information will exist and increase in the future. Not only in the old-fashioned printed way, with a tendency to decrease, but in a more ubiquitous, customized way and provided for free for many consumers (Lenzi 2017). To show the usefulness of the proposed framework model, described in Sect. 5, in this not so distant future, 4 (four) future newsroom scenarios were built, structured along three categories: Business model; Journalism professionals and Technology.

Applying the newsrooms structure in different scenarios using the proposed framework model. The production of news inside newsrooms is analyzed through the lens of 4 (four) different newsroom structures, evolving from Newsroom 1.0 to Newsroom 4.0 (Table 3).

8.1   Newsroom 1.0 Versus Scenarios A, B and C

This first feature of Newsroom 1.0, presented in Sect. 4, may coexist in the versions of the identified scenarios (A, B, and C). Here, the news may be printed and sometimes digitally accessible, through the institutional site or, indeed, through social networks.

Scenario A

A conventional news organization (big players) could be structured in four levels (operational, knowledge, management and strategic levels) and has many functional divisions (sales & marketing, financial & accounting, production and human resources divisions), as any organization of the industrial age. In this scenario, the heart of a news organization is the newsroom, which is analogous to the production division of a factory. The newsroom is where the newspaper content is produced and in a conventional newsroom the process of producing news in the knowledge/operational level is generally very structured and highly synchronized.

News production in Newsroom 1.0 has a traditional media ecosystem basically comprised of few sources and large communication companies. In this scenario (scenario A) the hierarchy is very rigid. However, the hierarchy has a minor role in small companies (scenario B) and in the communities of practices such as scientific communities (scenario C).

Scenario B

In scenario B, that of a local and regional newspaper network, the newsroom's management level may have many levels, and the most important one is the gatekeeper, who selects the news that will be printed or even published online. The newsroom's strategic level establishes the editorial line, regardless of the feature. With IT becoming increasingly inexpensive and telecommunications services accessible, journalists will be engaged in this scenario too.

As the available information grows at an exponential rate, it is a counter-productive process; moreover, it does not guarantee high quality news as a single version is produced through the sole gatekeeper who is responsible for the selection of published news and its

Table 3 Applying the model/framework in each possible scenario and structures

Newsroom version	Scenario A Largest corporations or big players	Scenario B Local and regional newspaper network	Scenario C Production of news for a specific community
Newsroom 1.0 Broadcast and centered on single newsroom	• Newsroom uses a proprietary CMS • The CMS is the core	• Newsroom uses free CMS software • However, it is hard and expensive to integrate free software with the WMS	• WMS: Slow process to evaluate and revise articles by peers • KMS: Knowledge is distributed among participants in the community of practice
Newsroom 2.0 Integrated and convergent Newsrooms	Newsroom could potentially use CMS, KMS and WMS	• Newsroom uses free CMS software • However, it is hard and expensive to integrate free software with semantics/open ontological approach • Nonetheless, one could use WMS to integrate processes • Relatively few success cases with KMS	• The newsroom has a WMS integrated with the CMS to add distributed spatial content development capabilities and capabilities to add new media to the content and showcase multiple channels (TV, radio, social net) • KMS: Could use semantics/open ontology • Editor will grant access to the CMS to improve productivity and quality
Newsroom 3.0 Distributed, Asynchronous, Digital printer, Curator Journalist with Semantic	• Newsroom has new types of sources and views to produce news • Uses CMS integrated with KMS and WMS • Uses Data Journalism, even for data generated by IoT devices and Robotics	• Newsroom has new possibilities of using the CMS and KMS with open ontologies • Possibility of using the WMS integrated with the CMS • Hardly uses Data Journalism, which is expensive for newsrooms of this size	• The editor that grants access to the CMS and KMS controls the WMS. In this context, editing an article becomes a dialogue with the research community. This dialogue improves quality and productivity • Uses Distributed Data Journalism, which exploits the distributed but collective intelligence and abilities of a community

(continued)

Table 3 (continued)

Newsroom version	Scenario A Largest corporations or big players	Scenario B Local and regional newspaper network	Scenario C Production of news for a specific community
Newsroom 4.0 The virtual newsroom is the social media	Proprietary Virtual Newsroom integrating CMS, KMS, WMS, Social Media and AI	Free Virtual Newsroom integrating CMS, KMS, WMS, Social Media and AI. Newsroom specialized in a few types of media or channel	Virtual Newsroom integrating CMS, KMS and WMS. Use of platforms for video and audio conferencing, chat and webinars, Skype, Google meeting, Zoom, Team-link, etc.

distribution to the readership. In this scenario, the team is in a place (newsroom) where people work together in a synchronous manner; therefore, this system needs to be strongly connected.

Scenario C

When one analyses the first line of Table 3, it is possible to understand how each scenario can adopt technologies for Newsroom 1.0 as well as the variations that occur regarding the use of the framework model's CMS, WMS and KMS. In this scenario, the operational level of the CMS is very intertwined with the relationship level of the WMS, and thus the newsrooms knowledge level of the KMS is hypertrophied.

8.2   Newsroom 2.0 Versus Scenarios A, B and C

Here we illustrate how each scenario (A, B and C) adopts features of Newsroom 2.0. It's no surprise to say that newspapers adopting the Newsroom 1.0 features are downsizing, and moving to Newsroom 2.0.

Scenario A

As the web evolved, new types of news media content were brought to the market: printed newspapers, radio, television, the Internet 2.0 and so on. Although the newsrooms were restructured to integrate the cited communication channels, the process to handle the new news media content, the basic production of content remains quasi similar to that of Newsroom 1.0. The production of narrative news almost always follows a synchronous process.

A concrete example of news organization where multiplatform integration is a reality in Scenario A is *La Nación* (de Mendonça et al. 2016). Another example is *O Globo* in 2015, where we identified many success cases of newsroom convergence and integration. Also, in this scenario, the data journalism occurs more frequently than in other scenarios, since it requires a team of specialists that only big media companies can afford. The loosely integrated and asynchronous cooperative environment can produce many versions of the same news, each one for a different media content type.

Scenario B

The integrated and convergent newsrooms of the small newspapers had to be restructured to handle many different types of media content and media sources, such as those stemming from news agency, big data, as well as to identify other confidential sources. The gatekeeper team guarantees the quality level of the news. As the team is in a place (newsroom) where people work together in synchronous ways, this modus operandi is still strongly connected. Also, smaller newspapers have an integrated newsroom to produce content to many channels with different communication languages (printed news, digital news, radio, TV, customized news (info services). In this Scenario (Scenario B), there are few divisions of the tasks and responsibility, and consequently some people can do various activities. Moreover, the media organizations use many technologies in their information systems (IS for marketing & sales, financial & accounting, human resources, etc.), which is implemented with different technologies. Consequently, they all demonstrate a loose integration.

Scenario C

The convergent newsrooms of Newsroom 2.0, presented in subsection 9.4.2 may coexist with production of news for a specific community. Normally, this type of newsrooms is small and they resolve the problems of quality well, since speed is not always a priority.

The use of semantic technologies can occur with the use of KMS to support a CMS or a WMS. The enormous quantity of articles produced in this scenario forces the use of semantic tools in order to enable search and research based on evidence. In this scenario, the production and edition of a specific article is a peer dialogue within the community, improving quality and productivity. Examining the row corresponding to Newsroom 2.0 of Table 3, it is possible to understand why Scenarios B and C can adopt a free or economical CMS.

8.3   Newsroom 3.0 Versus Scenarios A, B and C

Looking at the last row of Table 3, it is possible to understand, how the Newsroom 3.0 model can incorporate each scenario (A, B and C) in turn.

Scenario A

The evolution in the production of journalistic content, which comprises the retrieval of data and information, and the distribution of the news, goes through the said digital transformations (activities, processes, competencies and models to fully leverage the changes and opportunities). The said digital transformations[2] are already well practiced in several business segments and markets. The first challenge for New Journalism, sometimes called Cyber journalism, is to continue producing content with quality, in a scalable manner and with the dynamism and speed characteristic of the Internet. The said challenge must be achieved with minor costs in this Scenario A for big players (NY Times, The Guardian, Globo, NHK). To achieve these goals, many technologies are aggregated in the Newsroom

---

[2]The concept of digital transformation is present in the proposals of services of IT companies. The concept is currently understood as the intensive use of technology to radically improve the performance or reach of companies and governments. See Medeiros Neto, B. (2017). O Cidadão Contemporâneo frente às Tecnologias da Informação e Comunicação. São Paulo: Iglu Editora, p. 359.

3.0 model, in order to create a more dynamic newsroom virtual environment, where any journalist from any country could collaboratively contribute to produce any content at any time.

Scenario B

The second challenge is to solve the increase in the volume of news to be produced and the diversity of content (almost customized for each reader/client). In this scenario, we find the local and regional newspapers occupying this niche. However, news quality still remains a valuable asset, even if it is for short and objective news. Here, quality can be achieved through a curator (watch keeper) journalist, which is quite different to a gatekeeper: the former decides whether a given message will be distributed to the readers, whilst the latter decides what will be produced and published.

The activities related to data journalism have been more difficult to establish in this Scenario than the other two because small companies usually cannot afford a specialized team to undertake it.

Scenario C

The Newsroom 3.0 model applied to Scenario C is yet to be adapted, however this will be faster than in the other scenarios. The variations in the use of technology and semantic approaches occur in the same way as in the two previous scenarios, but with more intensity. The potential to use of a semantic CMS is not only strong here, but also a necessity, especially in the health/biology research area. The news quality based on ontologies stored in KMSs will be intensively used for scenarios A and C. Finally, in this Scenario C, the workflow for production of news (WMS) will be increasingly distributed.

8.4   Newsroom 4.0 Versus Scenarios A, B and C

Looking at the last row of Table 3 it is possible to understand how the Newsroom 4.0 model can incorporate each scenario (A, B, and C) in turn.

Scenario A

Here, the innovation will be a guarantee of the evolution of the production to virtual journalistic content, which comprises the retrieval of data and information in big data and verification of the news from distribution in social networks. The big companies have undergone changes, the said digital transformations. It was started in the Newsroom 3.0, for example, activities, processes, competencies, and models to fully leverage the changes and market opportunities will be completed.

The second challenge for New Journalism, sometimes called Cyber journalism, is to continue producing content with quality, in a scalable manner and with the dynamism and speed characteristic of the Internet, that is, with the proliferation of fake news. To achieve these goals, many technologies will be available and can be integrated into the Newsroom 4.0 model, where any journalist from any country, any time and using any access device could collaboratively contribute to producing any content at any time.

Scenario B

It depends on the knowledge and use of free tools, journalism of data, and IA. The possibilities to solve the increase in the volume of news to be produced, news verification and the diversity of content (customized for each reader/client) will lead to collaboration or dependence on the great newspapers of the Scenario A. In this Scenario, we find the local and regional newspapers and mural journalism agency, with growth potential. However, news quality and confirmability will be the main value of journalism, even if it is for short and objective news.

The quality and confirmability can be achieved by a specialist information system (intelligent agent), and through curators (watch keeper) or journalist collective. The former decides whether a piece of given news (information) will be distributed to the readers, whilst the latter decides what will be produced and published. The activities related to data journalism can, finally, establish in this Scenario, then the other two scenarios, because small companies usually cannot afford a specialized team to undertake it, but they will use a collective team.

Scenario C

The Newsroom 4.0 model applied to Scenario C is already happening due to the presence of social networks. The use of collaborative technology and semantic approaches occur in the same way as in the two previous scenarios, but after consolidating solutions. The potential to use of a semantic CMS will be a good practice, but also a necessity by access via the social network, especially in the health/biology research area, and environment monitoring. The news quality is supported on ontologies models applied on big data, where KMSs will be intensively used for scenarios A and C. Finally, in this Scenario C, the workflow for the production of news (WMS) will be increasingly distributed and collaborative.

9   Conclusion

Since the post-war era, the same business model has remained unchanged around the world, producing single media content by selling media products and advertisements, and spaces on media. The main feature was broadcast and its diffusion or the broadcast journalism. However, over the last three decades, the challenge has been to transform a traditional, stagnant production branch into an integrated and convergent environment. Today, many journals already this feature. At least for a good part of its productions changed to Convergent and Integrated Journalism. However, the arrival of the digital transformation is not only to replace an old or operational strategy of communication with a new, as consequence the mobile on the life of almost all people, but the introduction of a feature of new production of the news more flexible, innovation-friendly with use or support of Web 3.0.

We propose a framework model for newsrooms that captures both journalistic and technical considerations, and compared these features based on the 4C Collaborative Model. Moreover, our proposed framework doesn't forget the business model, the knowledge and skills of the journalism professionals and technology and proposes three prospective scenarios. The framework model that has been proposed in this paper has emanated as a result of visits to 5 newsrooms of highly respected journalistic organizations throughout the world. It is worth mentioning that the 4C model of collaborative newsrooms has not yet been applied to the context of convergent newsrooms, which is a task what we have undertaken in the present research. Throughout this paper it is possible to realize how perspectives of the Journalists content production from newspaper/print to Virtual Newsroom 4.0 will be in the future.

We have elaborated a framework model based on main facets, which translates to the dimensions of the newsrooms, in the present and the future: News Content Management (CMS); News Production Process (workflow); News Knowledge Management (KMS) and Social Media Management Systems (SMMS) all permeated with AI technology. It has been derived in the context of 3 scenarios, which have been devised as a result of our newsroom visits and deepen in the area of communication and information technologies.

The intention is for this framework model to give journalists and news companies the possibility to transcend the current business model with one in which the newsroom is immersed in the social media. We believe that the proposed newsroom model will help journalists, professionals and news organizations to increase their productivity, engage prosumers and also aid to produce high quality and timeous news. However, these are part of our future endeavors.

Acknowledgements This research has been sponsored by Fundação de Apoio à Pesquisa do Distrito Federal (FAP-DF), under the project "Flexible organizational information systems based on business processes with contextual semantic guidelines." Grant Number SEI 00193-00000096/2019-78. This case study of the project is part of the Experimental Laboratory for the Study of Digital Languages for Mobile Devices (Labdim) of the Department of Communication, University of Brasília, registered under number 485 707 at CNPq/2013-6, in partnership with the Department of Computer Science (UnB) and Brunel University London, UK.

References

Annane, A., Aussenac-Gilles, N., Kamel, M.: BBO: BPMN 2.0 Based ontology for business process representation. In Proceedings of ECKM 2019 20th European Conference on Knowledge Management, vol. 1, pp. 49–59. Lisbon, Portugal (2019)

Antoniou, G., Van Harmelen, F., Hoekstra, R.: A Semantic Web Primer, third edition. The MIT Press (2012)

Avilés, J.A.G.: Ocho gráficos que explican la transformación del periodismo. Retrieved from: https://medium.com/@jagaraviles/ocho-gr%C3%A1ficos-que-explican-la-transformaci%C3%B3n-del-periodismo-cd9a8abf4877 (2017). Access on 20 Apr 2019

Avilés, J.A.G., Prieto, M.C., Kaltenbrunner, A., Meier, K., Kraus, D.: Integración de redacciones en Austria, España y Alemania: modelos de convergencia de medios. Anàlisi. quaderns de comunicació i cultura **38**, 173–198 (2009)

Baños-Moreno, M.J., Felipe, E.R., Pastor-Sánchez, J.A., Lima, G.A.B.: Metadatos en noticias: un análisis internacional para la representación de contenidos en periódicos. In: XII Congreso ISKO España y II Congreso ISKO España-Portugal, November 19–20, Organización del conocimiento para sistemas de información abiertos. Murcia University, Murcia (2015)

Becker, V., Guimarães, E.M.: Adapte-se ou morra: Como The New York Times e Globo.com estão se moldando a um novo jornalismo. In: Nunes, P. (ed.) Jornalismo em Ambientes Multiplataforma, p 359 (2016)

Behrendt, W.: The interactive knowledge stack (IKS): a vision for the future of CMS. In: Maass, W., Kowatsch, T. (eds.) Semantic Technologies in Content Management Systems. Springer, Berlin, Heidelberg (2012)

Belochio, V.D.C.: Jornalismo em contexto de convergência: implicações da distribuição multiplataforma na ampliação dos contratos de comunicação dos dispositivos de *Zero Hora*. Doctoral dissertation. Postgraduate Program in Communication and Information, Federal University of Rio Grande do Sul, Brazil (2012)

Canavilhas, J.: Novos atores na redação: como muda o jornalismo? In: Martins, G.L., Reino, L.S. A., Bueno, T. (orgs.) Performance em ciberjornalismo. Tecnologia, inovação e eficiência. Ed. UFMS, Campo Grande, MS (2017)

Christensen, O.A., Jacobsen, V.K.: News Hunter: a semantic news aggregator. Retrieved from: http://bora.uib.no/handle/1956/16192 (2017). Accessed on 10 Dec 2017

Cook, N.: Enterprise 2.0: How Social Software will Change the Future of Work. Gower Publishing, Aldershot, England (2008)

Cordeiro, R.C., Lessa, W.D.: Designers' workflow and news visualization: the case of newspaper *O Globo*. Blucher Des Proc **1**(2), 1261–1269 (2014)

Costa, A.P., João Loureiro, M., Reis, L.P.: Do Modelo 3C de Colaboração ao Modelo 4C: Modelo de Análise de Processos de Desenvolvimento de Software Educativo. Revista Lusófona de Educação, vol 27 (2014)

Crilley, R., Gillespie, M.: What to do about social media? Politics, populism and journalism. Journalism **20**(1), 173–176 (2019) (Sage)

da Fonseca, M.B., Ishikawa, E., Neto, B.M., Victorino, M., Oliveira, E.C.: Ferramenta para Anotação Semântica de Processos de Negócio de uma Redação Jornalística (Tool for Semantic Annotation of Business Processes in a Newsroom). In: ONTOBRAS, pp. 239–244 (2018)

Dailey, L., Demo, L., Spillman, M.: The convergence continuum: a model for studying collaboration between media newsrooms. Atl. J. Commun. **13**(3), 150–168 (2005)

Davies, J.: The BBC is using 'slow news' to fight fake news. Available from https://digiday.com/uk/bbcs-slow-news-focuschanging-newsroom-dynamics/ (2017). Accessed on 17 Mar 2020

de Carvalho, J.M.: O conteúdo continua sendo o principal dilema das plataformas digitais. In: Martins, G.L., Reino, L.S.A., Bueno, T. (orgs.) Performance em ciberjornalismo. Tecnologia, inovação e eficiência. Ed. UFMS, Campo Grande, MS (2017)

de Deus, V.S., Ishikawa, E., Oliveira, E.C., Victorino, M., Neto, B.M., Groenli, T.M., Ghinea, G.: Towards a semantic-based content management system for journalistic writing. In: Proceedings of the 10th International Conference on Management of Digital Ecosystems, pp. 141–148. ACM (2018)

de Mendonça Jorge, T., Cardoso, S.G., Oliveira, E.C., Neto, B.M.: Experiencias de convergencia en Brasil y Costa Rica. Análisis del proceso de integración en redacciones periodísticas. Los casos de *Correio Braziliense*, *O Globo* y *La Nación* Convergence Experiences in Brazil and Costa Rica. Analysis of the integration process in newsrooms. The cases of *Correio Braziliense*, *O Globo* and *La Nación*. XIII Congreso de la Asociación Latinoamericana de Investigadores de Comunicación, Mexico City (2016)

Deuze, M.: Global journalism education: a conceptual approach. J. Stud. **7**(1), 19–34 (2006)

dos Santos, M.C.: Comunicação Digital e Jornalismo de Inserção. Labcom Digital, UFMA, São Luís (2016)

Dowd, C.: The scrabble of language towards persuasion: Changing behaviors in journalism. In: International Conference on Persuasive Technology, pp. 39–50. Springer, Berlin, Heidelberg (2013)

Ekström, M., Westlund, O.: The dislocation of news journalism: a conceptual framework for the study of epistemologies of digital journalism (2019)

Ermida, A.: Social Journalism: Exploring how Social Media is Shaping Journalism. The Handbook of Global Online Journalism. Wiley Online Library (2012)

Filippo, D., Pimentel, M., Wainer, J.: Metodologia de Pesquisa Científica em Sistemas Colaborativos. Sistemas Colaborativos **1**, 379–404 (2011)

Fricke, R., Thonsem, J.: http://community.mediamixer.eu/documents/ucnewsroom (2014). Accessed on 10 Dec 2017

Fuks, H., Raposo, A.B., Gerosa, M.A., Pimentel, M., Filippo, D., Lucena, C.D.: Teorias e modelos de colaboração. In: Sistemas Colaborativos, organized by Mariano Pimentel and Hugo Fuks, pp. 16–33. Elsevier, Rio de Janeiro (2011)

García, R.: Retrieved from: http://www.mediamixer.eu/webinars-on-semantic-newsroom-and-copyright-management-now-available/ (2014). Accessed on 10 Dec 2017

García, R., Perdrix, F., Gil, R.: Ontological infrastructure for a semantic newspaper. In: Semantic Web Annotations for Multimedia Workshop, SWAMM. Retrieved from: http://citeseerx.ist.psu.edu/viewdoc/summary?doi=10.1.1.97.9865 (2006). Accessed on 22 May 2015

Heller, B., Mello Junior, J.: As redes sociais e a edição de e-books [Social networks and e-book publishing]. Revista FAMECOS **24**(1), 1–18 (2017)

Hollingsworth, D.: Workflow Management Coalition: The Workflow Reference Model. Hampshire, U. K. Document Number TC00–1003, 19. Retrieved from: ftp://atenas.cpd.ufv.br/dpi/mestrado/Wkflow-BPM/The%20Workflow%20Reference%20Model.pdf (1995). Accessed on 17 Mar 2020

Huovelin, J., Gross, O., Solin, O., Linden, K., Maisala, S.P.T., Oittinen, T., Toivonen, H., Niemi, J., Silfverberg, M.: Software newsroom–an approach to automation of news search and editing. J Print Media Technol Res (2013)

IPTC.: NewsML standard. https://iptc.org/news/newsml-g2-2-29-released/ (2017)

Isaac, M.: Zuckerberg Plans to Integrate WhatsApp, Instagram and Facebook Messenger. New York Times. 25th January 2019. https://www.nytimes.com/2019/01/25/technology/facebook-instagram-whatsapp-messenger.html (2019). Access on 2 June 2019

Laje, N.: Ideologia e técnica da notícia. Série Jornalismo a Rigor.V.5., 148 p. Insular, Florianópolis (2012)

Larrondo, A., Domingo, D., Erdal, I.J., Masip, P., Van den Bulck, H.: Opportunities and limitations of newsroom convergence: a comparative study on European public service broadcasting organisations. J. Stud. **17**(3), 277–300 (2016)

Laudon, K., Laudon, J.: Management Information Systems: Managing the Digital Firm. Prentice Hall, New Jersey (2015)

Lenzi, A.: Inversão de papel: prioridade ao digital como um novo ciclo de inovação para jornais de origem impressa. PhD dissertation. Post-grad Program in Journalism, Federal University of Santa Catarina. Florianópolis (2017)

Maia, K.B.F., Agnez, L.F.: A convergência digital na produção da notícia: Dois modelos de integração entre meio impresso e digital. Proceedings of the 1st Colóquio Internacional Mudanças Estruturais no Jornalismo—(Mejor) (2011)

Medeiros Neto, B., Ghinea, G., Almeida, B.: Uso de workflow para avaliar a convergência tecnológica e prospectar a integração de meios e conteúdos nas redações: jornal *La Nación/* Costa Rica. VI Encontro da Ulepicc Brasil—GT3—Indústrias midiáticas. November 9–11, Brasília (2016)

Medeiros Neto, B., Ishikawa, E., Groen, T.-M., Ghinea, G.: Newsroom 3.0: Managing Technological and Media Convergence in Contemporary Newsrooms. Proceedings of the 52nd Hawaii International Conference on System Sciences, pp. 2407–2416 (2019)

Menke, M., Kinnebrock, S., Kretzschmar, S., Aichberger, I., Broersma, M., Hummel, R.Kirchhoff, S., Prandner, D., Salaverría, R.: Convergence culture in European Newsrooms: comparing editorial strategies for cross-media news production in six countries. J. Stud. 1–24 (2016)

Moreno-Schneider, J., Srivastava, A., Bourgonje, P., Wabnitz, D., Rehm, G.: Semantic Storytelling, Cross-lingual Event Detection and other Semantic Services for a Newsroom Content Curation Dashboard. Proceedings of the 2017 EMNLP Workshop on Natural Language Processing meets Journalism, pp. 68–73 (2017)

Oliveira, C.: Towards a new authoring environment: overview of some ontology based systems. In: ELPUB (2004)

Oliveira, E.C., Ishikawa, E., Horinouchi, L.H., Granja, T.H., de A Nunes, M.V., Rodriguez, D., Menegassi, R.B., Gois, L., Ghinea, G.: Designing an ontology based Zika virus news authoring environment for the Semantic Web. In: Proceedings of the 8th International Conference on Management of Digital Ecosystems, pp. 197–203

Palmonari, M., Uboldi, G., Cremaschi, M., Ciminieri, D. & Bianchi, F.: DaCENA: serendipitous news reading with data contexts. In European Semantic Web Conference, pp. 133–137. Springer International Publishing (2015)

Rublescki, A., Barichello, E.: Jornalismo colaborativo e redes sociais no mainstream: estudo comparado do jornal zerohora.com e do washingtonpost.com. Rumores, 7(14), 99–118 (2013)

Salaverría, R., Negredo, S.: Periodismo integrado: convergencia de medios y reorganización de redacciones. Sol90 (2008)

Salaverría, R., García-Avilés, J.A., Masip, P.: Concepto de convergencia periodística. In: García, X.L., Fariña, X.P. (coords.). Convergencia digital. Reconfiguración de los medios de comunicación en España. Santiago de Compostela: Servicio editorial de la Universidade de Santiago de Compostela, pp. 41–64 (2010)

Schauer, B., Zeiller, M.: E-Collaboration systems: how collaborative they really are. In: Proceedings of COLLA 2011–The First International Conference on Advanced Collaborative Networks, Systems and Applications. COLLA 2011–The First International Conference on Advanced Collaborative Networks, Systems and Applications (2011)

Schwingel, C.: Jornalismo convergente através de plataformas de altíssima resolução: o projeto 2014K. Jornalismo convergente: reflexões, apropriações, experiências, pp. 255–267. Insular, Florianópolis (2012)

Squirra, S.: As tecnologias mergulham a comunicação em uma cerebralidade artificial. In: Cibertecs: conceitos, interações automações futurações. Organizador S. Squirra. LabCom Digital, São Luís, MA (2016)

Temer, A.C.R.P., Nery, V.C.A.: Para entender as teorias da comunicação. Edufu, Uberlândia (2009)

Ternai, K., Török, M., Varga, K.: Corporate semantic business process management. In: Corporate Knowledge Discovery and Organizational Learning, pp. 33–57. Springer International Publishing (2016)

Thurman, N., Lewis, S.C., Kunert, J.: Algorithms, Automation, and News, Digital. Journalism 7(8), 980–992 (2019). https://doi.org/10.1080/21670811.2019.1685395

Troncy, R.: Bringing the IPTC news architecture into the semantic web. Semant Web-ISWC 2008, 483–498 (2008)

Undurraga, T.: Making News, Making the Economy: Technological Changes and Financial Pressures in B (2017)

von Rimscha, M.B., Verhoeven, M., Krebs, I., Sommer, C., Siegert, G.: Patterns of successful media production. Convergence, 1354856516678410 (2016)

Edison Ishikawa He is a professor of the Department of Computer Science of the Brasília University (UnB) since 2014. He received the Ph.D degree in Systems and Computer Engineering from the Federal University of Rio de Janeiro (COPPE/UFRJ) in 2003, the M.Sc. degree in Informatics from Pontifical Catholic University of Rio de Janeiro (PUC-RIO), the B.E. Degree in Computer Engineering from Military Institute of Engineering and the B.S. Degree from Agulhas Negras Military Academy. His current research interests focus on Semantic Computing, Systems of Systems Engineering, Distributed Systems and Security. At UnB he participates in research projects in those areas, with funding from Brazilian government agencies and institutions such as CNPq, FAP-DF and Brazilian Army. Participated in the implementation activities of the MDM Project: A model proposal for a semantic framework of a collaborative environment for information management in journalistic writing.

Benedito Medeiros Neto Post-Doctorate/Informatics: Semantic Framework for Journalism by CIC/IE/UnB (2018). Post-Doctorate: Digital Literacy and Mobile Learning by the School of Communication and Art/USP (2014). PhD in Information Science: Evaluation of Digital Inclusion programs, by FCI/UnB (2012). Master in Operational Research/Graph Theory by EST/UnB (1981). Specialist in Electrical Engineering/Artificial Intelligence by UnB (1986). Electrical/

Telecommunications Engineer by UnB (1975). Visiting Professor at Computer Science Department, Brunel University, London/UK, May 2018. Project Scholar/MEC/MCTI/CAPES/ CNPq/FAPs No. 09/2014. Researcher and Professor at UnB/IE/CIC and FAC/UnB. Associate Researcher at Escola do Futuro\USP (2014-). Consultant/Evaluator of FAPESB/BA. Reviewer at IGI Global. Associate of ASSOCIAÇÃO PROFISSÃO JORNALISTA (2019). Participant of the GT01/ENANCIB; SIMEDUC/UNIT/Aracaju; Ibero-American Magazine of CI/Faculty of Information Science/UnB. PROFESSIONAL LIFE: Director of Innovation and Development at IBrTec (2019-); At Ministry of Communications: Consultant for Digital Inclusion; Coordinator of Knowledge Management and Evaluation of the GESAC Program (2012). At ECT he was Director Manager (2002), Advisor to the Vice Presidency (1999), Advisor/Technical Support (FAT) to the Directorate of Technology and Infrastructure (1998) and Senior System Analyst (2007). He was Head of Telecommunications Section of the Telebrás System (1978). He was Professor at ESAP/ ECT (1988), Professor at CEUB/Brasília. DEVELOPMENT AND RESEARCH AREAS: Computer Science, Information and Communication; Network Engineering; ICT teaching; Informatics and Society; Collaborative Systems and Web; Semantic Web; Digital Inclusion; Digital Cities; Competence in Information, Social Networks and Evaluation of Innovation Programs. CNPq RESEARCH GROUPS: (a) JorTec/SBPJOR; (b) Journalism and Memory in Communication; (c) Technology and Digital Narratives); (d) Competence in Information.

Gheorghita (George) Ghinea is a Professor in Mulsemedia Computing in the Department of Computer Science, at Brunel University. Dr. Ghinea's research activities lie at the confluence of Computer Science, Media and Psychology. In particular, his work focuses on the area of perceptual multimedia quality and how one builds end-to-end communication systems incorporating user perceptual requirements. Dr. Ghinea has applied his expertise in areas such as eye-tracking, telemedicine, multi-modal interaction, and ubiquitous and mobile computing, leading a team of 8 researchers in these areas. He has over 300 publications in his research field and is the Editor in Chief of the International Journal of Pervasive Computing and Communications. Currently, his research pursuits are centered on extending the notion of multimedia with that of mulsemedia a term which he has put forward to denote multiple sensorial media, ie. media applications which engage three or more of the human sense. His work has been funded by both national and international funding agencies and has been covered by the BBC, Telegraph, and Forbes magazine, among others. He consults regularly for both public and private institutions in his areas of expertise.

APPENDIX: The MDM Project

Benedito Medeiros Neto and Edgard Costa Oliveira

Introduction

The Multimodal Digital Media Project (MDM Project) is a semantic computational model applied for convergent digital structures in journalistic newsrooms. It comprises an international study on information systems in newspaper offices in Brazil, Costa Rica, England, and the United States (in progress). The MDM Project began in February 2015, and was due to be concluded on January 31, 2018, as defined. It is a collaboration between professors of different departments of the University of Brasília (UnB): the Department of Computer Science (CIC) and the Graduate Program at the Department of Communication (FAC); subsequently, the Post-Graduation Program of the Department of Information Science (FCI) has also taken part in the research.

As part of the Brazilian Education Ministry program Ciência sem Fronteiras (Science without Borders), the MDM Project was funded by the Brazilian federal government and is in accordance with the guidelines of the Post-Doctoral Programs in Brazil (Capes/MEC and CNPq/MCTIC). The MDM Project focused on disseminating its model and results across the various media and research groups both in Brazil and overseas, putting the emphasis on publishing articles in national and international academic publications, and participating in relevant events in Brazil and abroad. The project still lacks pieces of information, this book is a response, but its most relevant results were disclosed

Benedito Medeiros Neto

Universidade de Brasília, 70910-000. Campus Darcy Ribeiro, Brasília, Brazil

e-mail: medeirosneto@unb.br

Edgard Costa Oliveira

Universidade de Brasília, 70910-000. Campus Darcy Ribeiro, Brasília, Brazil

e-mail: ecosta@unb.br

B. Medeiros Neto et al. (eds.), *Digital Convergence in Contemporary Newsrooms*, Studies in Systems, Decision and Control 370, https://doi.org/10.1007/978-3-030-74428-1

The MDM Project was based on theoretical references already established in the areas of Information Technology (IT), Journalism, Information, and Education, in order to shed some light on contemporary society's relationships. It also dealt with the use of Collaborative Systems in digital networks, and a journalistic work environment that incorporate fundamentals from Semantic Web, Ontologies, and Social Networks. The human being of the twenty-first century has new behaviors, and new ways of being and acting in a workspace, all of which is an easily observable truth in a convergent newsroom, where very intense human interactions take place.

Inportance of the MDM Project

The justification for preparing a book about the MDM Project is the fact that it involves a team of high-level studies, and reinforces national research groups beyond UnB, such as the federal universities of the Brazilian states of Maranhão and Espírito Santo. The three scholarships abroad, and three post-graduation programs at UnB's FAC, CIC, and FCI were boosted, and allowed new researchers to participate in the project. Scholars, professors, and students had the chance to live in a research environment, and some had the opportunity to visit and do internships at the University of Brunel, London, mainly in the last two years.

Also, the great variety of activities carried out by the MDM Project included seminars with students, professors, and invited guests; and connections to other institutions in Brazil and abroad, in order to promote the debate among the researchers. Researchers who addressed issues related to Changes in Newsrooms carried it out. The seminars were promoted in the format of open lectures to the academic community, research days, and courses for undergraduates. Courses in the three main areas were included in the project, with the participation of undergraduate and graduate professors and students. Part of the records of these scientific events will be preserved in the form of articles, and a book elaborated on well-defined criteria.

All members and participants of the MDM Project took into account its main goals:

(a) Specification and design of collaborative management and production environments for newsrooms, based on ontology and semantics;

(b) Proposition of a framework model that will unfold into a set of artifacts for journalistic production. On the educational side, it included the orientation of projects, and of monographs of students from CIC and FAC, and the rules or contributions in disciplines at CIC, FAC and FCI.

The scientific production, as presented at the end of this Appendix, includes the submission of scientific articles to national and international journals; publication of technical papers, and participation in national and international symposia and conferences in the areas of Computer Science, Information and Communication (Journalism).

Teaching, and R&D

Postdoctorals

As participants in the MDM Project, both teachers and researchers taught and worked together with their students, in addition to exchanging knowledge with other departments of the same university. The actions and activities that gave rise to the chapters of this book are based on a research project funded by the Brazilian federal government. The studies, mostly exploratory, resulted in research methodologies, case studies, scientific articles, and participation in conferences and seminars, and in the development of artifacts and applications in journalism. Scientific production is nearly always predominant in the fields of Computer Science, Information Science, and Communication.

This project departed from case studies of major newspapers, carried out according to empirical and investigative research, in the real context of newsrooms. The objectives and limitations of the project were the determining factors for the definition of the choice of places for visits, the context for the field survey, and its scope within each newsroom.

As a research method, we used the qualitative, exploratory, and descriptive approach, and also case studies, in order to elucidate the different points of view, values, and interpretations of individuals in their work group. In the second stage of the research, to evaluate the framework model, an intervention was planned in the newsroom of an experimental university newspaper, *Campus Multiplataforma*, at FAC/UnB. The model was implemented through the development of collaborative systems, via Web, supporting collaborative work in the newsroom.

The project offered three "sandwich"[1] scholarships abroad and three postdoctoral internships in Brazil. The researchers' presented the topics covered, as well as the details, in two reports approved by the Department of Computer Science:

(a) *Final Postdoctoral Report in Computer Science, Edgard Costa Oliveira, PhD, Faculty in UnB Gama—FGA Software Engineering*

---

[1]Brazilian nickname for a scholarship of one year abroad inserted in the middle of a doctoral or postdoctoral research.

This report presents the results of the postdoctoral research conducted by Edgard Costa Oliveira, enrollment FUB 1019864, professor of Software Engineering at the UnB Gama campus, in the context of the MDM Project, CAPES / PVE of the Science Without Borders Program.

The project was presented at Meeting 127 of the Collegiate of the Graduate Program in Informatics of CIC/UnB. This postdoctoral research was approved at that meeting, under the guidance of Prof. Maria Emília Walter, PhD.

The MDM Project number (Process: 88881.068354/2014) started in March 2015 and ended in January 2018. The project was proposed in partnership between CIC/UnB in partnership with the Communication Department of UnB (FAC), under the coordination of professors Maria de Fátima Brandão and Thais Jorge Mendonça, and with the collaboration of special visiting professor George Ghinea, from University of Brunel, London, who was the main supervisor of the research.

The candidate's working plan, whose results are presented in this report, is contained within the scope of the MDM Project, considering the following main goals:

- Case study of the needs of users of journalistic text production environments;
- Specification and prototype design of a journalistic newsrooms management environment based on semantics;
- Simulation of using the text production environment based on ontologies in a given journalistic context.

The proposed activities had salutary outcomes, such as the submission of scientific articles to conferences, publication of technical papers, orientation of Pibic projects, and discipline regulation at CIC/UnB.

(b)   *Final Postdoctoral Report in Informatics, Prof. Benedito Medeiros Neto, PhD, approved at Department of Computer Science (CIC/UnB), Institute of Exact Sciences (IE).*[2]

This final report is organized in three parts. The first deals with the presentation of the execution of the research project as a whole, based on its goals, and on the rationale and methodologies established in the propositional documents. The second part presents the main scientific production and the results achieved. It also contains a set of appendixes. And the third part is an individual assessment of the infrastructure, resources, research environment, and methodology for individual and group work. It started in February 2015; an opportunity in which my post-doctoral research in Brazil was being defined. Our participation as a collaborator

---

[2]Available at: https://www.researchgate.net/publication/327593817_INSTITUTO_DE_CIENCIAS_ EXATAS-IE_DEPARTAMENTO_DE_CIENCIA_DA_COMPUTACAO-CIC_Relatorio_Final_ de_Pos-Doutorado_em_Informatica

has taken place since the beginning, 2015, and we moved to the postdoctoral fellowship status in May 2017.

The post-doctoral research project intended to support a cooperative and collaborative work environment for journalistic production, first of all. The post doctorate research project intended to support a cooperative and collaborative work environment for journalistic production. The project identified technological solutions in suitable collaborative systems (groupware); workgroup techniques to support management; and the production of news stories in contemporary newsrooms. Technological solutions dealt with information virtualization, social networking, cloud computing, and new mobile devices (tablets and smartphones, but also other tangible and adaptable devices, such as ATAs). The project identified requirements and functionalities for each interaction model in terms of availability of information.

In research was used various software tools and information systems that enable users (journalists and professionals) to collaborate on both centralized or distributed tasks, used too by coordination of the information workflow, and in communication synchronously or asynchronously between people, now facilitated by the Internet. The growing incorporation of ICT in work environments favors cooperative work, particularly in mobile and Wi-Fi networks, urban environments, and newsrooms. At the same time, ICT is self are useful tools for living in contemporary society.

ICT facilitates communication between people and digital information processing systems, ensuring the coordination of people and material resources and cooperation in the work environment of journalistic production, to benefit reflective collective intelligence. Therefore, the bulk of this study was to determine whether semantic framework models in accordance with 4C Collaboration Models could be applied to facilitate journalistic work in the newsroom environment.

Center for Competence in Journalism and Technology

From the first semester of 2020 on, the Center focuses on teach-ing the four subjects, numbered below, at the Journalism Course/FAC, and also at the Department of Computer Science/IE. Actions, activities, and research resulting from the MDM Project nourish these subjects:

1. *Jornal Campus (FAC)*—Teacher: Zanei Barcelos.
   Partnership between "Campus Multimídia", a regular discipline of Journalism, and other activities for the development, implementation, and maintenance of a journalistic application (implementation of the application Campus Version 4.0), journalistic website (implementation of version 2.0), and Chatbot (implementation of version 1.0 and development of version 2.0) at the experimental laboratory newspaper *Campus Multiplatform*. Collaboration with the Computer Science College to receive some of its students on Special Topics in Journalism for direct work with Journalism students in the development of app and journalistic website.

2. Web Programming (CIC/IE). Teachers: Edison Ishikawa and Benedito Medeiros Neto. The "Advanced Topics on Computers—Prog Web" discipline provides students with the learning required for developing web applications, focusing mainly on using the Django framework with the Python 3.x programming language, Django extension with business process support (BPM), and extension to Semantic Web. Goals: at the end of the course, the student should be able to model and implement an information system in Django and Python, using the BPM and Web Semantics extensions.

The applications developed are in the area of Communication, more specifically Journalism, having as a real client the students of the Journalism course and the Department of Communication (FAC). In the last semester, four web systems were developed and continued this semester with developments and implementations, being three systems for the campus: app; local; and Chatbot, one for CEDOC.

Development of six systems, depending on the number of students enrolled. It comprises areas of experimentation and research with a focus on journalism. For instance: Education 3.0 (use of the web for teaching computers at the university, one system); Data Journalism (two systems for the use of tools, SPPS, R, spreadsheet, Python and Django programming for data collection on the web); Semantic Web and Ontology in Journalism (two systems); Artificial Intelligence and Deep Learning with fake news apps (one system).

3. Communication Topics—Data Journalism and Social Media (JOR/FAC). Teacher: Benedito Medeiros Neto; guests: journalists Larissa Silva and Thallita Alves Silva. The course aims at providing complementary knowledge in the area of computer science for social communication, notably journalism, enabling the updating and deepening of tools for searching and processing data in IT environments. It provides the student with the necessary skills to build content with the search for open data on the web or in open databases or available through the API. The course aims at training journalism students to use the tools to access and process Computer Science data in academic, personal, and professional life; and seeks to articulate the development of skills for the use and appropriation of ICTs with the development of collaborative work guided by project problems, to facilitate the interpretation of the data collected for the construction of quantitative content. Its specific objectives are:

   (a) To promote computational thinking and the use of tools and software for teaching, work and research, manipulation and transfer of data and documents;
   (b) To facilitate the creation and production of content based on data, texts, images, videos, and their publication in new electronic media;
   (c) To identify tools to support the development of research and data processing for the development of content for academic and professional purposes

Two web systems were developed by CIC/UnB students, which are now available for other students.

4. Communication, Information and Computing: fundamentals and application. Teachers: Rose May; Benedito Medeiros Neto; Monica Peres; Marcelo de Jesus, and Alzimar Ramalho (and other guests). The relationship between practice and theoretical foundations in the hybrid field of knowledge in the Communication, Information and Computer Sciences. The course comprises multicourse issues, solutions, and reflections that involve those areas in networked computational environments in computational environments. Basic concepts of Communication, Information and Computing for transdisciplinary and transversal application in research projects that involve communication and information in different types of networks in digital environments. The goals are to create a space for reflection, discussion, and sharing experiences and practices on:

(a) Skills in Communication, Information and Computing and the articulation of knowledge of this hybrid field with the Collaborative Knowledge Networks.

(b) Journalism and Organizational Communication as theoretical and instrumental support for planning communication and information actions in computational environments.

(c) Digital collection of different types of experimental products, and academic research, with the objective of preserving and disseminating the theoretical and practical knowledge of FAC.

5. The discipline promotes applied, interdisciplinary research, and is open to the participation of professors and undergraduate and graduate students, with the objective of observing empirical problems involving communication and network information in computational environments, as well as presenting solutions and transdisciplinary concepts. In this first semester of 2020, the research will take place also in a research laboratory in Memory and Information Management at CEDOC/FAC. At this point, the discipline involves finishing the digitalization of the collection of the newspaper *Campus*, created in 1970, and offering it in digital format at the Central Library of UnB. It is a pioneering initiative in practical training through laboratory newspapers in Brazil. The discipline's students will also catalog the collection of the Free Book project, and the monographs deposited at CEDOC. The development and implementation of the Information System and app for CEDOC/FAC (Version 2.0) will make accessible all FAC production.

Technical Visits

The survey of data and information on visits made by researchers are part of the MDM Project Collection. In this space were included recordings, interviews, websites, and electronic repositories available to everyone. Data analysis was carried out throughout the project. A brief summary of how technical visits took place can be seen below.

Visit to a Traditional Newspaper in Brasília

Correio Braziliense was the place where almost all members of the MDM Project had the opportunity to learn about the operation of a newspaper, and the production of actual news. It was possible to identify the main actors in the newsroom, and the tools for building a print newspaper and virtual news for the web and social networks. Visits to *Correio Braziliense* took place in 2015, 2016, and 2017. Those visits illustrated the main processes, and leveled the playing field for debates among researchers in the areas of Communication, Computing and Information, who entered the project during its development.

At Correio Braziliense, the meeting room has a panel that is used for agenda meetings of editors, daily, of the newspaper print, offering potential covers, for example. There are markers to show the goals achieved and whether the articles are good or not. Also, if was the exclusive matter of the Correio. This panel is just a sample of how long the interviews were, and always carefully followed by the re-searchers.

Publications related to the visit: Jorge, T. M.; Guedes, S; Costa, E; & Medeiros Neto, B (2016) Convergence experiences in Brazil and Costa Rica. Analysis of the integration process in newsrooms. The cases of *Correio Braziliense, O Globo,* and *La Nación.* XIII Congreso de la Asociación Latinoamericana de Investigadores de Comunicación, Mexico City, Mexico.

Technological Convergence and Integration at O Globo

With an average daily circulation of 183,000 copies, *O Globo* is the second Brazilian printed newspaper, right after the popular *Super Notícia* (220,971), and followed by *Folha de S. Paulo* (175,441), according to the *Associação Nacional de Jornais* (ANJ, the national newspaper association, 2015). In the convergence model of *O Globo,* journalists work for the core of news, regardless of the content's destination: paper, website, or other vehicles of the group. Like many communication companies, Group Globo recently launched popular newspapers, such as *Extra.*

During the researcher's visit, in 2015, it was observed that the model adopted in the integrated newsroom was partially convergent with the printed newspaper's editorial offices in charge of producing material for the print and digital versions (Maia and Agnez 2011). At the same visit to *O Globo,* managers showed researchers the transposition of news from the paper to the portals, something that happened little by little, and progressed over the next two years.[3]

Convergence and Integration at La Nación, Costa Rica

The vehicle integration process (the newspapers *La Nación*, *El Financiero*, *La Teja*, and monthly magazines) started in 2007, and had its first stage completed in 2011, with the move to an especially designed building. In January 2016, about 400 people were working in the newsroom. During the technical visit, *La Nación*'s journalists already worked under the concept of "digital first".

The case study of *La Nación*'s newsroom, which employed 440 professionals (270 journalists, and 170 IT, design, and specialists from other areas), realized that the first attempt at integration and relevant convergence, in 2007, faced strong resistance to the use of IT.[4]

Multimodal Digital Media (MDM) at BBC

The BBC newsroom was a remarkable visit for the researchers of the project. Also called "The World's Newsroom," it is well located at the center of London. More than just looks, the newsroom reflects a merger of BBC international journalism with the state of art in technology. The space was created to privilege multimedia production for multiple platforms: TV, radio, and online and virtual communication for various parts of the world.[5]

The BBC's first investments to integrate the content in a semantic way occurred in 2007 with the creation of the BBC Programs (database with the indexation of all BBC radio and television programs), and in 2010, based on the W3C consortium. The BBC's goal was "building coherence media", according to the interviewed professionals: the user should be able to find everything the BBC has published on

---

[3]See Medeiros Neto, B. (November 2017). Uso del workflow para evaluar la convergencia tecnológica y el alcance de la integración de multimedios y contenidos en las redacciones: periódico *O Globo*, Río de Janeiro. Conference: ICOM 2017, November 13–17. La Habana, Cuba: Palacio de Convenciones.

[4]See Medeiros Neto, B., Almeida, B. & Ghinea, G. (November 2016). Uso de workflow para avaliar a convergência tecnológica e prospectar a integração de meios e conteúdos nas redações: jornal *La Nación*/Costa Rica. Conference: VI Encontro Nacional da União Latina da Economia Política da Informação, da Comunicação e da Cultura (ULEPICC).

[5]See Santos, É., Medeiros, B., Lenzi, A., & Ghinea, G. (2019). Journalistic newsrooms in a context of convergence: a comparative exploratory study in Brazil, Costa Rica and England. Communication & Innovation, 20 (43). Available at: https://www.researchgate.net/publication/335128966_Redacoes_jornalisticas_em_contexto_de_convergencia_um_estudo_comparativo_exploratorio_no_Brasil_na_Costa_Rica_e_na_In Inglaterra_NEWSROOMS_IN_A_CONTEXT_OFAN_VER_VER.

its website on a particular subject, following a particular semantic line. For example, reading a news story about a musician and then browsing the TV shows that had him as a guest artist. "The Cooperation Dimension Using Frameworks, and Collaborative Systems in the Newsroom—A Case Study of the BBC" is an article about the visit awaiting publication.

Reuters and Its Digital Multimedia News

News agencies are companies specialized in distributing data and news directly from the sources of an event to media outlets, such as newspapers, magazines, radio stations, websites, and television stations. The first agencies appeared in the middle of the nineteenth century, with the Havas Agency (today France-Presse), founded by the writer and translator Charles-Louis Havas, in 1835.

Reuters, the news and media division of Thomson Reuters, is the world's largest international multimedia news provider, and was named number one international brand for digital reach in an Ipsos Affluent Europe 2017 survey, ahead of 44 international media brands on digital platforms. It reaches more than a billion people every day. Reuters provides business, financial, national and international news to professional media market clients worldwide.

The idea of a convergent newsroom was not strange to the Executive Director, who expressed enthusiasm for changes that could make the environment more organic and rational, which should occur a month after the visit. As an international news agency, Reuters cultivates the tradition of working together, and has the honor of not making mistakes, which means that a permanent copy staff (proofreaders) is preserved, when most vehicles decreed the extinction of this function. Reuters works with 14,000 employees, 2500 of who are journalists operating in 204 cities, and broadcasting news in 19 languages.[6]

Global Evaluation of Research Works

This evaluation seeks to verify the achievement of some objectives of this research project. Besides UnB, other Brazilian federal universities developed high-level studies, and national research groups have been strengthened in some of them, such as at the federal universities of Maranhão and Espírito Santo. The three Graduate Programs of FAC, CIC, and FCI at UnB were boosted and allowed new researchers

---

[6]See Franciscato, C. E., Martins, E., Jorge, T. M., Neto, B. M., Martins, G. L., Werdemberg, A., ... & Bueno, J. V. (2017, September). Inovações no Jornalismo. Mesa Coordenada da Rede Jortec. In 15° Encontro da SBPJor. Available at: https://www.researchgate.net/publication/321010360_Redacoes_Jornalisticas_-_convergencia_e_inovacao_em_tres_continentes_SBPJor_2017.

and students to participate in the research. MDM Project fellows, professors, and students had the opportunity to live in a real research environment, with some of them visiting and working at Brunel University, in London, mainly in the last two years of the project. Some of the hurdles were:

(a) Selection of doctoral and postdoctoral fellows, especially at end of the project;
(b) Setting up research environments at CIC, FAC, and FCI;
(c) Overcoming relationship issues within the project team, many of them to be expected in an interdisciplinary group.

But the MDM Project also benefited by other issues:

(a) The project, as it addresses a relevant and very current subject, has always aroused interest from students and teachers. We had a constant flow of people entering and leaving the project. This had a positive side because many people became aware of the public policy to encourage Science.
(b) The Special Visiting Professor, George Ghinea, from London's Brunel University, acted in a competent and skillful way to overcome the main difficulties that arose throughout the project, such as: a very heterogeneous team; research areas with slightly different methods; and differentiated forms of work in the areas of Communication, Information, and Computing. On the other hand, it must be said that all his visits to Brasília were impactful, and brought substantial gains to the work, contributing to overcoming the hindrances that were emerging within the project.

Evaluation of the Project Infrastructure

Physical and Material Resources

The volume of resources spent on funding the activities met just what was expected for the MDM Project. Many requests from participants were not responded to because they were not anticipated in the original budget, or because of lack of the funding resources approved at the time. The allocation of financial resources was thus rigorous due to CAPES/CNPq's rules, which at times prevented actions that could bring immediate results to the project. Many items foreseen in the budget did not take place, because they did not bring expected gains or were not necessary.

Systems and Environments

The MDM Project lasted three years, during which we had less productive moments, and other moments that were more productive, mainly at the end, when the scholarship holders were incorporated into the project's actions. It is certain, though, that each group used the best software to facilitate their research tasks. The systems for Brunel's CIC, FAC, and Computer Department environments were chosen. The development of experiments was more restricted to people who were dealing directly with the research object in the laboratories, at the time. There was no single environment used by everyone, only document records and reporting were accessible.

In the end, the technical and academic coordination were more committed to giving access to all participants to the results and artifacts of the project. In a way, it was possible to make contents available to students and teachers at UnB. It was part of the tasks of the members of the project to share the results in classrooms, via disciplines, conferences, and national and international events, and this has been happening successfully.

Working Methods

The knowledge of management tools and agile working methods by various components of the project, in a way, added little to the development of teamwork. Most of the time, traditional development methods were adopted, even in spite of several attempts to use new methods in the weekly meetings of the project monitoring Thursdays, and within the groups. As a consequence, production was good at times and reasonable at other times.

Results Achieved

Research for newsroom collaborative work environments mediated by Information Technology (IT), and its implementation, attract software service and product companies, academic institutions and competence centers. Therefore, the present research had the potential to contribute scientifically to the area, and to give visibility to postgraduate programs, in particular in the UnB's departments of Computer Science (CIC), Communication (FAC), and Information Science (FCI), warranted by the participation of the Special Visiting Researcher.

The postdoctoral internships researches supported the MDM Project. Two researchers did technical visits to the Department of Computer Science at Brunel University, in London, as beneficiaries of "sandwich" doctoral scholarships, and three post-doctoral researchers received scholarships in Brazilian programs to conduct the project's activities in the absence of the visiting professor.

The MDM Project enhanced the performance of researchers in Teaching, Research, and Extension, having most of the activities carried out in the Laboratory of Special Projects of CIC/IE, at in the University of Brasília, in partnership with its Graduate   Program.   The   university's   experimental   newspaper   *Campus*

Multiplataforma continues to be a space for the application of artifacts, such as the development of systems based on the Tetrahedral Model, that is, a semantic framework of a collaborative journalistic environment.

Other relevant results were postdoctoral internships, development of doctoral dissertations, publication of scientific articles, participation in national and international events by UnB and Brunel University's researchers. In a way, this strengthened the partnership with the Laboratory of Experimentation in Digital Languages for Mobile Devices (Labdim) at FAC. (registered with CNPq under number 485707/2013-6).

Other Scientific Production

Assunção, A.B.M., Jorge, T.M.: #…-As mídias sociais como tecnologias de si. Esferas, (5) (2015)

Brandão, M.F.R., Oliveira, E.C., deMendonça Jorge, T., Cardoso, S.G., Neto, B.M., Ghinea, G.: Convergent digital media in journalistic newsrooms: an ecosystem perspective (2015). https://www.researchgate.net/profile/Benedito_Medeiros_Neto/publication/325930239_Cyberculture_and_Journalism_An_Age_of_Coolaboration_and_Connection/links/5b2d43354585150d23c5eddb/Cyberculture-and-Journalism-An-Age-of-Coolaboration-and-Connection.pdf

daFonseca, M., Ishikawa, E., Medeiros Neto, B., Victorino, M., Oliveira, E.C.: Ferramenta para Anotação Semântica de Processos de Negócio de uma Redação Jornalística (Tool for Semantic Annotation of Business processes in a Newsroom). In: ONTOBRAS, pp. 239–244 (2018)

de Deus, V.S., Ishikawa, E., Oliveira, E.C., Victorino, M., Neto, B.M., Groenli, T.M., Ghinea, G.: Towards a semantic-based content management system for journalistic writing. In: Proceedings of the 10th International Conference on Management of Digital Ecosystems, pp. 141–148 (2018, September)

Ishikawa, E., Medeiros Neto, B.: Newsroom 3.1: Incorporating Social Media Management in Semantic Newsrooms with Flexible Business Process. Escola do Futuro, ECA\USP. São Paulo, November 11-14 (2019). http://seminariohispano-brasileiro.org.es/ocs/index.php/viishb/viiishbusp/paper/view/602

Maia, K.B.F., Agnez, L.F.: A convergência digital na produção da notícia: Dois modelos de integração entre meio impresso e digital (2011). http://www.mejor.com.br/index.php/mejor2011/MEJOR/paper/view/73

Marques, M., Medeiros Neto, B., Peres, M.R.: Comunicação, Informação e Computação: experiências interdisciplinares no ensino, pesquisa e extensão (2017)

Medeiros-Neto, B.: Monitorando a passagem do Jornalismo online ao Webjornalismo, marcado pela cultura profissional dos professores e pesquisadores. Conference: 8° Congresso Internacional de Ciberjornalismo - Campo Grande – MS. At: Campo Grande – MS (2017)

Medeiros-Neto, B., Ishikawa, E., Barcelos, Z.R., de Deus, V.S.: Modelo para um framework semântico de ambiente trabalho colaborativo para gestão da informação em redação jornalística. SBPJor Conference: 16° Encontro Nacional de Pesquisadores em Jornalismo FIAM-FAAM/Anhembi Morumbi. São Paulo, November (2018)

Molina, F., Medeiros, B.: O perfil do jornalista 3.0: Novas competências necessárias para o jornalismo no século XXI. Simpósio Internacional de Educação e Comunicação (SIMEDUC), (8) (2017)

Neto, B.M.: Os telefones celulares e a aprendizagem colaborativa na sociedade de serviços: o desdobramento da sociedade da informação em rede. P2P e Inovação, **3**(2), 96–123 (2017)

© The Editor(s) (if applicable) and The Author(s), under exclusive license to Springer Nature Switzerland AG 2022
B. Medeiros Neto et al. (eds.), *Digital Convergence in Contemporary Newsrooms*, Studies in Systems, Decision and Control 370, https://doi.org/10.1007/978-3-030-74428-1

Neto, B.M., Ishikawa, E., Ghinea, G., Grønli, T.M.: Newsroom 3.0: managing technological and media convergence in contemporary newsrooms. In: Proceedings of the 52nd Hawaii International Conference on System Sciences (2019, January)

Oliveira, E.C., Ishikawa, E., Horinouchi, L.H., Granja, T.H., de Nunes, A.M.V., Rodriguez, D., Ghinea, G.: Designing an ontology-based Zika virus news authoring environment for the semantic web. In: Proceedings of the 8th International Conference on Management of Digital Ecosystems, pp. 197–203 (2016a, November)

Oliveira, E.C., Ishikawa, E., Ghinea, G., Brandão, M.D.F.R., Victorino, M., Horinouchi, L.H.: Relato de experiência no ensino de web semântica e ontologias em cursos de graduação em Engenharia de Software e Ciência da Computação da UnB. In: ONTOBRAS, pp. 199–203 (2016b)

Oliveira, E.C., Ishikawa, E., Granja, T.H., de Almeida Nunes, M.V., Hironouchi, L.H., de Souza, C.C., Gois, L.: Ontology-based Zika Virus News Authoring Environment for the Semantic Web. In: ONTOBRAS, pp. 187–198 (2016c)

Oliveira, E.C., Ishikawa, E., Brandão, M.D.F.R., Victorino, M., Horinouchi, L.H.: O uso de software e plataformas no ensino de Web Semântica e ontologias na graduação em engenharia de sotware e da computação: relato de caso na UnB. ONTOBRÁS 2016. Curutiba/PR (2016d)

Oliveira, E.C., Ishikawa, E., Hironouchi, L.H., Granja, T.H., Marcos, V.D.A., Rodriguez, D., Ghinea, G.: Ontology-based CMS news authoring environment. In: 2017 IEEE 11th International Conference on Semantic Computing (ICSC), pp. 264–265). IEEE (2017)

Resende, V.G., Medeiros Neto, B., deDeus, V.S.: Ontologias como suporte a um CMS na elaboração de matérias jornalísticas do Campus Online da FAC/UnB. Conference: 17º Encontro Nacional de Pesquisadores em Jornalismo organized by SBPJor (Brazilian association of researchers on journalism). November 6-8. Federal University of Goiás (UFG) (2019)

Santos, É., Medeiros, B., Lenzi, A., Ghinea, G.: Redações jornalísticas em contexto de convergência: um estudo comparativo exploratório no Brasil, na Costa Rica e na Inglaterra. Comunicação & Inovação, **20**(43) (2019)

CPSIA information can be obtained
at www.ICGtesting.com
Printed in the USA
LVHW080605091122
732583LV00007B/246

9 783030 744304